Circular Economy in the Construction Industry

The Circular Economy in Sustainable Solid and Liquid Waste Management

Series Editor:
Dr. Sadhan Kumar Ghosh,
Professor, Mechanical Engineering, & Chief Coordinator, Centre for Sustainable Development and Resource Efficiency Management, Jadavpur University, Kolkata, India
President, International Society of Waste Management, Air and Water (ISWMAW)

Biomethane through Resource Circularity: Research, Technology and Practices
Edited by Sadhan Kumar Ghosh, Michael Nelles, H.N. Chanakya, and Debendra Chandra Baruah

The Circular Economy in Construction Industry
Sadhan Kumar Ghosh, Sannidhya Kumar Ghosh, Benu Gopal Mohapatra, and Ronald L. Mersky

For more information about this series, please visit:
https://www.routledge.com/The-Circular-Economy-in-Sustainable-Solid-and-Liquid-Waste-Management/book-series/CESSLWM

Circular Economy in the Construction Industry

Edited by
Sadhan Kumar Ghosh
Sannidhya Kumar Ghosh
Benu Gopal Mohapatra
Ronald L. Mersky

CRC Press is an imprint of the
Taylor & Francis Group, an **informa** business

First edition published 2022
by CRC Press
6000 Broken Sound Parkway NW, Suite 300, Boca Raton, FL 33487-2742

and by CRC Press
2 Park Square, Milton Park, Abingdon, Oxon, OX14 4RN

© 2022 selection and editorial matter, Sadhan Kumar Ghosh, Sannidhya Kumar Ghosh, Benu Gopal Mohapatra, Ronald Mersky; individual chapters, the contributors.

CRC Press is an imprint of Taylor & Francis Group, LLC

Reasonable efforts have been made to publish reliable data and information, but the author and publisher cannot assume responsibility for the validity of all materials or the consequences of their use. The authors and publishers have attempted to trace the copyright holders of all material reproduced in this publication and apologize to copyright holders if permission to publish in this form has not been obtained. If any copyright material has not been acknowledged please write and let us know so we may rectify in any future reprint.

Except as permitted under U.S. Copyright Law, no part of this book may be reprinted, reproduced, transmitted, or utilized in any form by any electronic, mechanical, or other means, now known or hereafter invented, including photocopying, microfilming, and recording, or in any information storage or retrieval system, without written permission from the publishers.

For permission to photocopy or use material electronically from this work, access www.copyright.com or contact the Copyright Clearance Center, Inc. (CCC), 222 Rosewood Drive, Danvers, MA 01923, 978-750-8400. For works that are not available on CCC please contact mpkbookspermissions@tandf.co.uk

Trademark notice: Product or corporate names may be trademarks or registered trademarks and are used only for identification and explanation without intent to infringe.

Library of Congress Cataloging-in-Publication Data
Names: Ghosh, Sadhan Kumar, editor. | Ghosh, Sannidhya Kumar, editor. |
    Mohapatra, B. C., editor. | Mersky, Ronald L. (Ronald Lee), editor.
Title: Circular economy in construction industry / edited by Sadhan Kumar
    Ghosh (professor, mechanical engineering, & chief coordinator, Centre
    for Sustainable Development and Resource Efficiency Management, Jadavpur
    University, Kolkata 700032, India), Sannidhya Kumar Ghosh (researcher,
    structural engineering and structural mechanics, Department of Civil,
    Environmental and Architectural Engineering, University of Colorado
    Boulder, USA), Benu Gopal Mohapatra (professor, civil engg & director,
    Consultancy Services, KIIT University, Bhubaneswar, India), Ronald
    Mersky (associate professor & chair, Department of Civil Engineering,
    director of solid waste project, director of engineering outreach,
    Widener University, Chester, PA 19013, USA).
Other titles: Circular economy in the construction industry
Description: First edition. | Boca Raton : CRC Press, 2022. | Series: The
    circular economy in sustainable solid and liquid waste management |
    Includes bibliographical references and index.
Identifiers: LCCN 2021033668 (print) | LCCN 2021033669 (ebook) | ISBN
    9781032108964 (hbk) | ISBN 9781032108971 (pbk) | ISBN 9781003217619
    (ebk)
Subjects: LCSH: Sustainable construction. | Construction industry--Waste
    minimizaton. | Construction and demolition debris--Recycling.
Classification: LCC TH880 .C57 2022 (print) | LCC TH880 (ebook) | DDC
    690.028/6--dc23
LC record available at https://lccn.loc.gov/2021033668
LC ebook record available at https://lccn.loc.gov/2021033669

ISBN: 978-1-032-10896-4 (hbk)
ISBN: 978-1-032-10897-1 (pbk)
ISBN: 978-1-003-21761-9 (ebk)

DOI: 10.1201/9781003217619

Typeset in Times
by SPi Technologies India Pvt Ltd (Straive)

Contents

Foreword vii
Preface... ix
Acknowledgements... xi
Editors xiii
Contributors xv

Part I   Sustainable Construction Practices and Circular Economy............................ 1

1. **Circular Economy in Combating Construction and Demolition Wastes Including Seismic Debris** 3
   Sannidhya Kumar Ghosh

2. **Circular Economy in Construction: An Overview with Examples from Materials Research**.. 11
   Dhanada K Mishra and Jing Yu

3. **Use of Industrial Waste Slag in the Development of Self-Compacting Concrete for Sustainable Infrastructures**.. 23
   K.P. Sethy, K. K. Sahoo, and Biswajit Jena

4. **Influence of Functionally Graded Region in Ground Granulated Blast Furnace Slag (GGBS) Layered Composite Concrete**............................... 29
   Sangram K. Sahoo, Benu Gopal Mohapatra, Sanjaya Kumar Patro, and Prasanna K. Acharya

5. **Utilization of Fly Ash as a Replacement of Sand in Concrete for Sustainable Construction**.. 37
   Ankita Sikder and Sukamal Kanta Ghosh

6. **Properties of Concrete at Elevated Temperature Using Waste HDPE as Fibre and Copper Slag as Mineral Admixture**... 43
   Marabathina Maheswara Rao and Sanjaya Kumar Patro

7. **Utilization of Geo-Waste in Production of Geo-Fiber Papercrete Bricks** 53
   Areej Palekar, Suraj Patil, and Uma Kale

8. **Effect of Slag Addition on Compressive Strength and Microstructural Features of Fly Ash Based Geopolymer** 61
   Dipankar Das, Alok Prasad Das, and Prasanta Kumar Rout

9. **Impacts of Municipal Solid Waste Heavy Metals on Soil Quality: A Case of Visakhapatnam**.. 69
   P.V.V. Prasada Rao and G. Siva Praveena

10. **Effective Utilization of Industry Solid Waste into the Concrete and Its Management**........... 77
    V.S. Vairagade, B.V. Bahoria, Rakesh Patel, P.T. Dhorabe, V.R. Agrawal, and N.P. Mungle

11. **Utilization of Industrial Waste in Normal Concrete: A Review** 87
    Srishti Saha, Tribikram Mohanty, and Bitanjaya Das

12. **Greenhouse Effect by Investigating an Internal Combustion Engine (IC Engine) Using Argemone Mexicana (Waste Plant) Biodiesel Blends**.. 95
    Akshaya Kumar Rout, M.K. Parida, and Mamuni Arya

v

13. Fertiliser Plant Phosphogypsum: Potential Applications in Agriculture and Road Construction 103

Durgasi Hariprasad, Harish Chandra Singh, Pranab Bhattacharyya, and Ranjit Singh Chugh

Part II Waste Utilization and Soil Stabilization 111

14. Bearing Capacity of Reinforced Soil on Varying Footing Size 113

Bandita Paikaroy, Sarat Kumar Das, and Benu Gopal Mohapatra

15. Improvement of Properties of an Expansive Soil with Induction of Bacteria 119

Suresh Gunji, C. H. Sudha Rani, C.H. Paramageetham, Yugandhar Gundlapalli, and Basha Sreenivasulu

16. Application of Treated Mixed Fruit Wastes in Soil Stabilization 129

David O. Olukanni and Chukwume G. Ijeh

17. Development of Flexible Pavement Cost Models for Weak Subgrade Stabilized with Fly Ash and Lime 139

Satya Ranjan Samal and Malaya Mohanty

18. Use of Fly Ash and Lime for Attainment of CNS Properties in a Swelling Soil 145

K.V.N. Laxma Naik, A. Thanusree, and C. H. Sudha Rani

19. Interface Shear Strengths between Bagasse Ash and Geogrid 153

Aditya Kumar Bhoi, Jnanendra Nath Mandal, and Ashish Juneja

Part III Sustainable Green Concrete 163

20. Experimental Investigation on Geopolymer Concrete with Low Density Aggregate 165

Sanghamitra Jena, Ramakanta Panigrahi, and Subrat Kumar Padhy

21. Strength Development in Ferrochrome Ash-Based Geopolymer Concrete 175

Jyotirmoy Mishra, S. K. Das, Bharadwaj Nanda, Sanjaya Kumar Patro, and Syed M. Mustakim

22. Investigation on Strength Factor of Composite Concrete Using Quarry Dust and Artificial Aggregates 185

Sudheer Ponnada, Partheepan Ganesan, and G. Praveen

23. Clean C&D Waste Material Cycles through BIM-enhanced Building Stock Examination Practices: An Austrian Case Study 191

M. Rašković, A.M. Ragoßnig, K. Kondracki, and M. Ragoßnig-Angst

24. Fly Ash-Based Jute Fiber Reinforced Concrete: A Go Green Approach for the Concrete Industry 199

Bidhan Ghosh and T. Senthil Vadivel

Part IV Energy Recovery and Resource Circulation in Construction 207

25. A Study on Tensile Strength and Modulus Properties of Concrete Using Industrial Waste Vermiculite and Granite-Fines 209

Damodhara Reddy Budda, Kiran Kumar Narasimhan, Jyothy S. Aruna, and C. Sasidhar

26. Eco-Friendly Utilization of Industrial Sludge as a Building Material: A Study of Steel Industries in the Tarapur Region, Maharashtra 219

Tarang Jobanputra, Vaidik Gajera, and Gaurav Kapse

Index 229

Foreword

The World Green Building Council has issued a bold new vision for how buildings and infrastructure around the world can reach 40% less embodied carbon emissions by 2030 and achieve 100% net zero emissions, building by 2050 to *net zero operational carbon goals. The majority of the world's population lives in cities, projected to rise to 70% by 2050. As cities continue to grow, and temperatures continue to rise dangerously, it has never been more important for the buildings and construction sector to be leading the way on climate action. Because the sector is responsible for such a large chunk of global emissions, it means there is huge potential for reduction. It is excellent news that the pathways laid out by the World Green Building Council contain an interim target for 2030—as we know how important it now is for getting the world on track to limit global temperatures to 1.5°C. Now the construction industry sector needs to mobilize immediately to put these changes into action for global benefit and encourage resource recovery initiatives for achieving this transformation.* Embodied carbon emissions have been overlooked in the past but, as shown by milestone research from the Intergovernmental Panel on Climate Change (IPCC), achieving drastic cuts in all carbon emissions over the next decade is required. Addressing upfront carbon is therefore crucial to fighting the climate crisis, as new construction is expected to double the world's building stock by 2060 causing an increase in the carbon emissions occurring right now.

Mankind has been following a linear model of production and consumption since the industrial revolution. Researchers report that global material use has tripled over the past four decades, with annual global extraction of materials growing from 22 billion tonnes in 1970 to 70 billion tonnes in 2010. The latest report from the International Resource Panel, Global Material Flows and Resource Productivity, provides a comprehensive, scientific overview of this important issue. It shows a great disparity of material consumption per capita between developing and developed countries. This has tremendous implications for achieving the Sustainable Development Goals (SDGs) in the next 11 years. Global material use has been accelerating. Material extraction per capita increased from 7 to 10 tonnes from 1970 to 2010, indicating improvements in the material standard of living in many parts of the world. It has been observed that 100.6 billion tonnes of materials were consumed in 2017, the latest year for which data is available while only 9.1% is circular, leaving a huge circularity gap and the scope of enhancing resource efficiency. Half of this total materials consumed is sand, clay, gravel and cement used for building, along with the other minerals quarried to produce fertiliser. Coal, oil and gas make up 15% and metal ores 10%. The final quarter are the plants and trees used for food and fuel. Raw materials have been transformed into goods that are afterward sold, used and turned into waste that has been many times unconsciously discarded and mismanaged. On the other hand, the circular economy is an industrial model that is regenerative by intention. One of the goals of the circular economy is to have a positive effect on the planet's ecosystems and to stop the excessive exploitation of natural resources.

The circular economy has the potential to reduce greenhouse gas emissions (GHGs) and the use of raw materials, optimize agricultural productivity, decrease the negative externalities brought by the linear model and enhance resource efficiency and productivity. The circular economy is said to be a system that aims to eliminate waste and minimize the consumption of resources, which paves the way for a sustainable world. It involves production models like sharing, remanufacturing, reuse, recycling, renovation, and repair, to create a closed loop to not only minimize the use of resources but also to reduce the creation of waste, pollution and carbon emissions. In this system, waste from a particular industry becomes the raw material of another or as a regenerative resource for nature. The regenerative approach of a circular economy is found in contrast with the traditional linear economy that involves production model like take, make, and dispose or extract-produce-consume-dispose-deplete (epcd2). The circular economy is advocated to be more and more sustainable than the linear economy. Reducing the use of resources and encouraging the use of waste materials helps in the conservation of resources and reduction of environmental pollution. Circular economy applied to the construction industry is of utmost importance to

vii

enhance the resource efficiency and effective construction business. In this context the book, Circular Economy in the Construction Industry, is a very important document for all the stakeholders.

9 December 2021

Sadhan Kumar Ghosh
Jadavpur University, Kolkata, India

Sannidhya Kumar Ghosh
University of Colorado Boulder, USA

Benu Gopal Mohapatra
KIIT Deemed to be University,
Bhubaneswar, India

Ronald L. Mersky,
Widener University, Chester, USA

Preface

This proposed book entitled, Circular Economy in the Construction Industry, will present some research and case studies that encourage reduction of the carbon emissions. Building and construction are responsible for 39% of all carbon emissions in the world, with operational emissions (from energy used to heat, cool and light buildings) accounting for 28%. The remaining 11% comes from embodied carbon emissions or 'upfront' carbon that is associated with materials and construction processes throughout the whole building lifecycle. To fully decarbonize, the sector requires eliminating both operational and embodied carbon emissions. This book aims to present the research, theories, policies and real-life case studies that utilize various types of waste materials in various fields of construction industries to gain an effective re-circulation for resource recovery and implementation of circular economy concepts thus cost-effective resource conservation. The chapters of this book are divided into four parts:

1. Sustainable construction practices and circular economy
2. Waste utilization and soil stabilization
3. Sustainable green concrete
4. Energy recovery and resource circulation in construction

Part I contains the chapters on the sustainable use of waste materials like ground granulated blast furnace slag, fly ash, copper slag, plastic waste and municipal solid waste as different ingredients of concrete. Some innovative concrete grades such as geopolymer concrete and functionally graded concrete are produced using some of the aforesaid waste materials and innovative materials for seismic and energy retrofitting of the existing buildings using C&D wwastes. These chapters explain sustainable construction practices leading to the circular economy.

Part II encloses chapters on stabilization and improvement of soil characteristics in some poor nature soils using waste materials like fly ash, lime dust, bagasse ash, and mixed fruit waste materials. Improvement in bearing capacity, interface shear strength, etc., are reported in the chapters.

Part III brackets chapters on sustainable green concrete using waste materials such as quarry dust, construction and demolition waste, fly ash, ferrochrome ash, etc., in preparation of environmentally compatible green concrete. These types of concretes are reported to be economy and ecology friendly.

Part IV encapsulates chapters on the use of rejects such as mine tailings, granite fines, industrial sludge, phosphogypsum, etc. These materials are reported to be used sustainably as a building material, road-making material, and concrete making material which shows energy recovery from waste and resource circulation.

Circular economy leading to the sustainable management of waste materials in construction industries is a very significant topic of interest. This book contains chapters presenting the latest research and case studies on reuse, recycle and energy recovery of a large number of waste stream materials applied to the construction industry. These waste materials are fly ash, ferrochrome ash, rice husk ash, bagasse ash, copper slag, mine tailing, geo waste, phosphogypsum, ground granulated blast furnace slag, chopped fibers, high-density polyethylene, industrial sludge, low-density aggregate, recycled aggregate, red mud, etc. The 4 parts of this book cover the reuse of waste stream materials which pave the way for sustainable concepts and enhancement of the circular economy. Since waste utilization, is the hot topic of the era, each the chapters will help filling the knowledge gap to some extent. The focus on the improvement of soil characteristics using waste materials, which explores many new ideas to treat the poor nature of the soils and make them fit for construction. The green concrete made with waste materials is a very important input for researchers and industries, which has also been dealt with in the book. Waste materials are used as binder and aggregate. The use of such materials as the binder will reduce the production of cement and therefore reduce carbon emission. The use of these materials as an aggregate will conserve natural resources. The use of rejected materials and their utilization will reduce the burden of the environment and release valuable landmass. The results obtained from these research papers open new areas of research and implementation inputs that will enhance the concept of the circular economy in related areas of concern.

Research works presented in the book show how waste stream materials can successfully be used in various fields in construction industries to achieve resource circulation and resource efficiency. This book will help in encouraging the implementation of circular economy in construction industries and establishing more green buildings following green building codes of using recycled construction materials and cost-effective construction with quality.

The ideas shared in this book are appropriate to fill the knowledge gap for the readers and researchers in general. Moreover, books dealing with the circular economy in construction industries are scarcely available in the market. This book is expected to be popular among the researchers, implementing agencies, entrepreneurs, service providers of C&D waste business, municipal administration tackling Construction and Demolition wastes related issues for resource recovery, academic and research institutes, and the policymakers. For all these reasons the proposed book, *Circular Economy in the Construction Industry* is a good collection for the libraries in academic institutions, government, and industry.

9 December 2021

Sadhan Kumar Ghosh
Jadavpur University, Kolkata, India

Sannidhya Kumar Ghosh
University of Colorado Boulder, USA

Benu Gopal Mohapatra
KIIT Deemed to be University,
Bhubaneswar, India

Ronald L. Mersky
Widener University, Chester, USA

Acknowledgements

The editors acknowledge the support of the chairman and the organizing committee of 8th IconSWM 2018, 9th IconSWM-CE 2019 and 10th IconSWM-CE 2020 that was attended by 44 countries and the governing council of the International Society of Waste Management, Air and Water for allowing the authors to contribute the articles for this book as well as the contributing authors.

The editors acknowledge the support and encouragement of the following organizations with gratitude.

- Members in the Secretariat of the International Society of Waste Management, Air and Water (ISWMAW). and Consortium of Researchers in International Collaboration (CRIC).
- UNCRD, Japan and IPLA, Global Secretariat at ISWMAW.
- Representatives joined from UNEP, UNDP, World Bank in IconSWM-CE.
- Researchers in the, INDIA H20 Horizon 2020 research project supported by DBT, Govt of India and European Union.
- The Ocean Plastic Turned into an Opportunity in Circular Economy (OPTOCE)-Project funded by SINTEF, Norway in India, Vietnam, China, Myanmar and Thailand
- The Indo-Hungary Industrial research Project at mechanical engineering department, Jadavpur University supported by DST, Govt of India and Hungarian Government
- The GCRF ESRC, UK funded research project on Social Wellbeing and Mental Health issues of employees in SMEs in India, Thailand, Bangladesh and the UK
- Researchers in Royal Academy of Engineering (RAE), UK funded research project on Circular Economy Adoption in SMEs in India and the UK.
- Circular Economy, social wellbeing, and mental health issues in SMEs.
- Members in the Centre for sustainable Development and Resource Efficiency Management, Jadavpur University.
- Members of Quality Management Consultants.
- The editors express their gratitude to the individuals whose support and encouragement are thankfully acknowledged.

Editors

Sadhan Kumar Ghosh is Professor in Mechanical Engineering & Chief Coordinator, Centre for Sustainable Development and Resource Efficiency Management at Jadavpur University, India. He also served as the Dean, Faculty of Engineering and Technology and Head of Mechanical Engineering. He was the Director, CBWE, Ministry of Labor and Employment, Govt. of India, and at Larsen & Toubro Ltd. He is a renowned personality in fields on waste management, circular economy, green manufacturing, supply chain management, sustainable development, co-processing of hazardous & MSW in cement kilns, plastics waste & e-waste management & recycling, management system standards (ISO), and total quality management (TQM), having three patents approved. Professor Ghosh is the founder Chairman of the IconSWM; President, International Society of Waste Management, Air and Water (ISWMAW) and the Chairman, Consortium of Researchers in International Collaboration (CRIC). He has received several awards in India and abroad including the distinguished visiting fellowship by the Royal Academy of Engineering, UK to work on 'Energy Recovery from MSW'. He has written 9 books, 40 edited volumes, more than 200 national and international articles and book chapters and was Associate Editor of special issues of WM&R and IJMCWM. His significant contribution has been able to place the name of Jadavpur University in the world map of research on waste management. He is consultant and international expert of UNCRD)/DESA, Asian Productivity Organization (APO), Japan, China Productivity Council (CPC), SACEP Sri Lanka, IGES Japan, etc. His international research funding include, EU Horizon 2020, Erasmus plus, UKIERI, Royal Society–DST, GCRF UK, Royal Academy of Engineering, Georgia Govt., etc. He is the leader of the Collaborative International Research Project on *"Global Status of Implementation of Circular Economy* (2018-2022)" by ISWMAW involving experts from 35 countries. He was the convener of ISO TC 61 WG2, member in the Indian mirror committee of ISO TC 207 & ISO TC 275. He is expert committee member of government initiatives and was the State Level Advisory Committee Member of Plastics Waste (Management & Handling) Rules 2011, expert committee member for the Preparation of standards for RDF for utilization set up by the Ministry of Housing and Urban Affairs (MoHUA), Govt. of India, Chair Elect of 11th IconSWM-CE & IPLA Global Forum 2021 to be held during December, 1–4, 2021. He is available at: sadhankghosh9@gmail.com & www.sadhankghosh.com.

Sannidhya Kumar Ghosh BCE (Hons), MS (USA), PhD (USA), researcher and acting as Instructor of Structural Analysis in Structural Engineering & Structural Mechanics, Department of Civil, Environmental and Architectural Engineering at The University of Colorado Boulder, USA. His research areas include studying seismic properties of aged concrete (funded by The National Institute of Standards and Technology (NIST) USA), concrete creep, concrete design, building design in a high seismic region, design of fluid viscous dampers for seismic vulnerability mitigation of structures, structural dynamics, construction and demolition waste management, etc. He did his graduation in civil engineering from Jadavpur University, India and MS from University of Colorado Boulder, USA, having one year of industry experience. He carried out projects at Aston Business School, Aston University, Birmingham and Loughborough University UK under the UKIERI Scheme, funded by the British Council and at Indian Institute of Technology, Kanpur under TEQIP funding. He has book chapters in American Society of Civil Engineers and a few publications. He is the governing body member of International Society of Waste Management, Air and Water and Research Chairman's Secretariat of 11th IconSWM-CE 2021 & member of Consortium of Researchers in International Collaboration (CRIC).

Benu Gopal Mohapatra PhD (IIT, Bombay), Professor and Director, Consultancy Services in KIIT Deemed to be University, Bhubaneswar & acted as the Director of School of Civil Engineering. His research focus includes Geotechnical Engineering related to pile foundation, ground improvement, soil structure interaction, and slope stability analysis. With 10 years of industry experience in Atkins Global Ltd., the UK, L&T(ECC)—in Dubai and India; Delhi Underground Metro Corridor—MC1B and Fugro Geotech Ltd., he has successfully delivered several foundation/geotechnical solutions as lead designer for Crossrail London Rail Ltd, London Underground, Network Rail, UK, M50 Slope Widening Options, Ireland, etc. He is actively involved in delivering various critical consultancy projects on landslide mitigation using advanced technologies for National Highway, NHIDCL, NHAI etc. He serves as a member of Institute of Civil Engineers UK, British Geotechnical Society, Fellow Member, Institution of Engineers & Indian Geotechnical Society. He served as the Secretary of 9th IconSWM-CE 2019.

Ronald L. Mersky PhD, PE, BCEE, Chair & Professor of the Department of Civil Engineering and Director of Solid Waste Project, Director of Engineering Outreach, at Widener University, He is Editor of The Journal of Solid Waste Technology and Management and Chair of The International Conference on Solid Waste Technology and Management. Dr. Mersky lectures extensively worldwide on solid waste, sustainability, and public communication of environmental issues. He also serves as a consultant and advisor to private industry and governments. He has authored a significant number of articles in international journals and books.

Contributors

Prasanna K. Acharya
School of Civil Engineering
KIIT Deemed to be University
Bhubaneswar, Odisha, India

V.R. Agrawal
Department of Civil Engineering
Priyadarshini College of Engineering
Nagpur, Maharashtra, India

Jyothy S. Aruna
Civil Engineering Department
S.V. University College of Engineering
Tirupati, Andhra Pradesh, India

Mamuni Arya
Radhakrishna Institute of Technology and
    Engineering, B.P.U.T.
Bhubaneswar, Odisha, India

B.V. Bahoria
Department of Civil Engineering
Priyadarshini College of Engineering
Nagpur, Maharashtra, India

Pranab Bhattacharyya
Navratna R&D Centre
Paradeep Phosphates LTD
Paradeep, Jagatsinghpur, Odisha

Aditya Kumar Bhoi
Department of Civil Engineering
IIT Bombay
Mumbai, India

Damodhara Reddy Budda
S.V. College of Engineering and Technology
Andhra Pradesh, India

C. H. Paramageetham
Department of Microbiology
Sri Venkateswara University
Tirupati, Andhra Pradesh, India

Ranjit Singh Chugh
Navratna R&D Centre
Paradeep Phosphates LTD
Paradeep, Jagatsinghpur, Odisha, India

Bitanjaya Das
School of civil Engineering
KIIT Deemed to be University
Bhubaneswar, Odisha, India

Dipankar Das
Department of Material Science and Engineering
Tripura University (A Central University)
Suryamaninagar, Agartala, Tripura, India

S. K. Das
Gron Tek Concrete & Research
Khordha, Odisha, India

P.T. Dhorabe
Department of Civil Engineering
Priyadarshini College of Engineering
Nagpur, Maharashtra, India

Vaidik Gajera
U.V. Patel College of Engineering
Bapunagar, Ahmedabad, India

Partheepan Ganesan
Department of Civil Engineering
MVGR College of Engineering (Autonomous)
Chintalavalasa, Vizianagaram, Andhra Pradesh, India

Bidhan Ghosh
Department of Civil Engineering
School of Engineering & Technology, Adamas
    University
Jagannathpur, Kolkata, India

Sukamal Kanta Ghosh
Greater Kolkata College of Engineering and
    Management
West Bengal, India

Sannidhya Kumar Ghosh
Dept. of Civil Environmental, Architectural
   Engineering
University of Colorado Boulder
Boulder, CO, USA

Yugandhar Gundlapalli
Department of Civil Engineering
Sri Venkateswara University
Tirupati, Andhra Pradesh, India

Suresh Gunji
Department of Civil Engineering
Sri Venkateswara University
Tirupati, Andhra Pradesh, India

Durgasi Hariprasad
Navratna R&D Centre
Paradeep Phosphates Limited
Paradeep, Jagatsinghpur, Odisha, India

Chukwume G. Ijeh
Department of Civil Engineering
Covenant University
Canaan land, Ota, Nigeria

Biswajit Jena
Department of Civil Engineering
DRIEMS (Auto.)
Cuttack, Odisha, India

Sanghamitra Jena
Department of Civil Engineering
Veer Surendra Sai University of Technology,
   VSSUT, BURLA
Sambalpur, Odisha, India

Tarang Jobanputra
Charotar University of Science and Technology
Veraval, Gujarat

Ashish Juneja
Department of Civil Engineering
IIT Bombay
Mumbai, India

Uma Kale
Environmental, M.H. Saboo Siddik College
   of Engineering
Mumbai University
Mumbai, Maharashtra, India

Gaurav Kapse
Manubhai Shivabhai Patel Department of Civil
   Engineering
Charotar University of Science and Technology
Anand, Gujarat, India

K. Kondracki
Vermessung Angst ZT GmbH
Vienna, Austria

Sarat Kumar Das
I.I.T(ISM)
Dhanbad, India

Kiran Kumar Narasimhan
Civil Engineering Department
S.V. University College of Engineering Tirupati
Tirupati, Andhra Pradesh, India

Subrat Kumar Padhy
Department of Civil Engineering
Veer Surendra Sai University of Technology,
   VSSUT, BURLA
Sambalpur, Odisha, India

Sanjaya Kumar Patro
Department of Civil Engineering
Veer Surendra sai University of Technology
Burla, Odisha, India

Akshaya Kumar Rout
KIIT Deemed to be University
Bhubaneswar, India

Prasanta Kumar Rout
Department of Material Science and Engineering
Tripura University (A Central University)
Suryamaninagar, Agartala, Tripura, India

K.V.N. Laxma Naik
Department of Civil Engineering
Sri Venkateswara University
Tirupati, Andhra Pradesh, India

Jnanendra Nath Mandal
Department of Civil Engineering
IIT Bombay
Mumbai, India

Dhanada K Mishra
KMBB College of Engineering and Technology
   (BPUT)
Odisha, India

Contributors

Jyotirmoy Mishra
Veer Surendra Sai University of Technology
Magurgadia, College Road
Keonjhar, Odisha

Tribikram Mohanty
School of Civil Engineering
KIIT Deemed to Be University
Bhubaneswar, Odisha, India

Benu Gopal Mohapatra
School of Civil Engineering
KIIT Deemed to be University
Bhubaneswar, Odisha, India

Marabathina Maheswara Rao
Veer Surendra sai University of Technology
Burla, Odisha, India

Malaya Mohanty
School of Civil Engineering
KIIT Deemed to be University
Bhubaneswar, Odisha, India

N.P. Mungle
Department of Civil Engineering
Priyadarshini College of Engineering
Nagpur, Maharashtra, India

Syed M. Mustakim
E&S Department
CSIR-IMMT, Acharya Vihar
Bhubaneswar, Odisha, India

Bharadwaj Nanda
Veer Surendra Sai University of Technology
Sambalpur, Odisha, India

David O. Olukanni
Department of Civil Engineering
Covenant University
Canaan land, Ota, Nigeria

Bandita Paikaroy
School of Civil Engineering
KIIT Deemed to be University
Bhubaneswar, Odisha, India

Areej Palekar
M.H. Saboo Siddik College of Engineering
Mumbai University
Mumbai, Maharashtra, India

Ramakanta Panigrahi
Department of Civil Engineering
Veer Surendra Sai University of Technology,
    VSSUT, BURLA
Sambalpur, Odisha, India

M.K. Parida
Department of Mechanical Engineering
C.V. Raman College of Engineering, B.P.U.T.
Bhubaneswar, Odisha, India

Rakesh Patel
Department of Civil Engineering
Priyadarshini College of Engineering
Nagpur, Maharashtra, India

Suraj Patil
M.H. Saboo Siddik College of Engineering
Mumbai University
Mumbai, India

Alok Prasad Das
Department of Life Science
Rama Devi Women's University
Odisha, India

P.V.V. Prasada Rao
Department of Environmental Sciences
Andhra University
Visakhapatnam, Andhra Pradesh, India

G. Praveen
Department of Civil Engineering
MVGR College of Engineering (Autonomous)
Chintalavalasa, Vizianagaram, Andhra Pradesh,
    India

A.M. Ragoßnig
RM Umweltkonsulenten ZT GmbH
Vienna, Austria

M. Ragoßnig-Angst
Vermessung Angst ZT GmbH
Vienna, Austria

M. Rašković
RM Umweltkonsulenten ZT GmbH
Vienna, Austria

Srishti Saha
School of Civil Engineering
KIIT Deemed to be University
Bhubaneswar, Odisha, India

K. K. Sahoo
Kalinga Institute of Industrial Technology
Bhubaneswar, India

Sangram K Sahoo
School of Civil Engineering
KIIT Deemed to be University
Bhubaneswar, Odisha, India

Satya Ranjan Samal
School of Civil Engineering
KIIT Deemed to be University
Bhubaneswar, Odisha, India

C. Sasidhar
Department of Civil Engineering
JNTU
Anantapur, Andhra Pradesh, India

T. Senthil Vadivel
Department of Civil Engineering, School of
   Engineering & Technology
Adamas University
Jagannathpur (PO), Kolkata

K.P. Sethy
Government College of Engineering Kalahandi
Bhawanipatna, India

Ankita Sikder
KIIT Deemed to be University
Bhubaneswar, Odisha, India
Sikder Bari, K.B.M.
Chakdaha, Nadia, Calcutta

Harish Chandra Singh
Navratna R&D Centre
Paradeep Phosphates LTD
Paradeep, Jagatsinghpur, Odisha, India

G. Siva Praveena
Department of Environmental Sciences
Andhra University
Visakhapatnam, Andhra Pradesh, India

Basha Sreenivasulu
Department of Microbiology
Sri Venkateswara University
Tirupati, Andhra Pradesh, India

C. H. Sudha Rani
Department of Civil Engineering
Sri Venkateswara University
Tirupati, Andhra Pradesh, India

Sudheer Ponnada
Department of Civil Engineering
MVGR College of Engineering (Autonomous)
Chintalavalasa, Vizianagaram, Andhra Pradesh,
   India

A. Thanusree
Department of Civil Engineering
Sri Venkateswara University
Tirupati, Andhra Pradesh, India

V.S. Vairagade
Department of Civil Engineering
Priyadarshini College of Engineering
Nagpur, Maharashtra, India

Jing Yu
School of Civil Engineering
Sun Yat-Sen University
Guangzhou, PR China

Part I

Sustainable Construction Practices and Circular Economy

1

Circular Economy in Combating Construction and Demolition Wastes Including Seismic Debris

Sannidhya Kumar Ghosh
University of Colorado Boulder, Boulder, USA

CONTENTS

1.1 Introduction 3
1.2 Construction and Demolition Waste (C&D) Generation Including Disasters Debris 4
1.3 Management and Recycling 6
1.4 C&DW Legislation 7
1.5 Discussion, Analysis, and Conclusion 8
Acknowledgement 8
References 8

1.1 Introduction

Disasters occur in various forms creating lots of problems for mankind and for the environment. The modes of disaster may happen in many forms, such as, natural or man-made; sudden onset (such as earthquake, fire, flood, tsunami, hurricane and volcano) or prolonged onset (such as civil conflict or drought); with varying degrees and types of physical and social impacts: The impacts of disaster in terms of material damage exceed the capacity for self-recovery of local communities. Disasters in many events create significant volumes of debris. In many cases, this satiation exceeds the existing city capacity of waste management. More than 3.0 billion tonnes annually, up to 2012, of construction and demolition waste (C&DW) was generated in 40 countries with an increasing trend. This C&DW generation has a different dynamic than the generation of municipal waste. C&DW can be generating from manmade activities as well as from natural disasters. Hence the generation of C&DW in many countries does not match with the estimated value. Nepal is one of such examples.

Sand and gravel are the main constituents: 35 tons of non-metallic minerals was extracted globally in 2010. A decrease in the quantity of raw materials and effective long-term planning can enable materials in structures to work as storage banks for significant materials and items following the concepts of 3Rs (Reuse/Recover/Recycle) and circular economy in the afterlife of the building. The waste products in the construction industry account for 30% of landfill waste worldwide (Ghaffar et al., 2020). India generates an estimated 150 million tonnes of C&DW every year (Building Material Promotion Council) while the recorded recycling capacity is 6,500 tonnes per day (Vidyasekar, 2019). This needs the support from government and respective legislation in the country. This study reviews the status of generation and utilisation of C&DW including disaster wastes. Policy instruments exist in a few countries, such as, India, USA, and a several other countries, management, and recycling. A case study from Nepal is presented. The study presents a generic model of C&DW management generated from manmade and natural disaster paths. The existing 6–9% of the global economy that is circular needs to be enhanced.

Eco-conscious construction trends and the increasing number of non-residential green buildings in the USA are driving the explosion in demand for green building materials while the green building materials market is estimated to reach $1 trillion in 2021. Non-residential buildings compose 40–48% of green

DOI: 10.1201/9781003217619-2

buildings in the USA, which was only 1.4% in 2005. This green construction trend was responsible for over 3.3 million jobs created in 2018.The Global Green Building Materials market was estimated at $199.9 billion. The US Green Building Council 2019 report states that, LEED (Leadership in Energy and Environmental Design)—certified homes have grown 19% since 2017 and are at an all-time high with nearly 0.5 million LEED-certified housing units globally and more than 0.4 million units located in the USA. The Canadian green building sector can reduce the Greenhouse Gases (GHG) by 53 megatons in comparison to 2018 levels. This will enhance the interest and business for green building materials in the construction of greenhouses [Canada Green Building Council (CaGBC)]. Due to the pandemic Covid-19, the market throughout the world was negatively impacted. However, this is a temporary setback that is expected to recover soon in the coming years.

1.2 Construction and Demolition Waste (C&D) Generation Including Disasters Debris

When new buildings and civil-engineering structures, such as highways and streets, bridges, utility plants, piers, and dams, are built, renovated or demolished generation of C&DW materials occur. The rate of generation of C&DW is increasing with the population increase and the GDP growth in many countries. These C&D materials consists of bulky & heavy materials, such as, concrete, wood, asphalt, gypsum, metals, bricks, glass, plastics, doors, windows, plumbing fixtures, trees, stumps, earth, rock from clearing sites, and many others. C&DW is one of the largest waste supply chain varying from 30–40% of total solid waste in large-scale C&D activities due to the accelerated urbanisation and city rebuilding (Akhtar & Sarmah, 2018; Jin et al., 2017; Zhao et al., 2010). Globally, about 35% of the C&DW produced is directed to landfills, without any further treatment, although efforts to recycle and reusing C&DW are increasing day by day (Maria et al., 2018). The growth rate of C&DW generation is quite higher in Australia (44%) (Poon et al., 2013). The rates of urbanisation in Bangladesh, between 2000 and 2010, was the fastest among the South Asian countries with an increase of 28% in 2011 (World Bank, 2015). Nearly 3.0 billion tons of C&DW annually, until 2012, was generated in 40 countries with an increasing trend.

There are two generation paths of C&DW—manmade C&DW and natural disaster debris. India generates an estimated 150 million tons of C&DW annually. It has been observed that the amount of C&DW generated varies from 30–200% of the municipal waste generated. Data management of C&DW is very weak in most of the developing countries. Researchers worked on quantification and optimisation models of C&DW (Ghosh et al., 2016, Alcántara-Ayala 2002). C&DW can be classified into three categories taking the phases into consideration: construction waste, renovation waste and demolition waste. Typical components in C&DW comprising of, bricks, wood, steel & metals, plastics, glass and a few others are reused and recycled traditionally (Ghosh & Ghosh, 2011; 2016a, b). Nearly 600 million tons of C&D debris, nearly two times the municipal solid waste (MSW) generation, were generated in the USA in 2018, as per Environmental Protection Agency (EPA) report. Quantification and measurement of C&DW recycling is very difficult as the end materials are handled and recycled in different ways and following different supply chain. As a part of measuring total solid waste (TSW) diversion, C&D recycling is surveyed in the USA. In Philadelphia, Delaware, it has been reported in a Statewide Waste Management (SWM) (Statewide SWM Plan, 2020). Plan that the estimated amount of C&DW disposed on an annual basis essentially remained constant over the past ten years: in the year 2008, the estimated C&DW disposed was 312,000 tons, compared with 316,000 tons in 2018, through there is a significant increase in generation. This estimate does not include the small amounts of C&DW that are disposed by mixed in with residential and commercial MSW (2020 SWM Plan).

Researchers show increasing interest in different aspects of C&DW as the importance of C&DW is increasing for the resource conservation and resource circulation. The summary of a few selected recent research and their results are presented in Table 1.1.

Manmade construction and demolition as well as the natural disaster result in debris generation. The type and severity of the disaster and the nature of the built environment dictate the composition and amount of the waste generation. Recent natural disasters such as the devastating Nepal earthquake on 25 April 2015 (Guha-Sapir et al., 2012; ICIMOD 2010; Karunasena et al., 2009) preceded a major aftershock

TABLE 1.1

Findings and Conclusion of Selected Research on Various Aspects of C&DW

Findings and Conclusions in the Literature	References
This study reviewed on the current status of generation, handling and important steps undertaken to manage C&DW in India. The study concluded that 3Rs strategy and technologies, e.g., design for deconstruction or reuse of materials should be adopted for reducing C&DW.	Mayur Shirish Jain (2021)
This study signposts a future research direction. It shows the requirements of reverse logistics in C&DW management, and to assist policymakers for developing informed policies to reduce negative environmental impact or related activities.	Brandão et al. (2021)
This research presents the recent trends in the on-site planning, logistics, performance and overall management of C&DW. It also focusses on standardisation practices and legislation. It concluded that new schemes for C&DW management in general tend to ensure the quality of the final product only considering that other aspects are available assuming these to be less expensive than treatments for other wastes.	Gálvez-Martos and Istrate (2020)
This paper studied the society-specific applicability of waste and related policies. It concludes that the system of building demolition management practices in Nigeria, could be more sustainable for the developing countries than that which is obtainable in the rich countries.	Aminu Lawan and Angela (2020)
This research involves onsite data collection from a dismantled building in Kolkata consisting of 16 flats with 2 rooms, 1 kitchen and 1 toilet. Major wastes generated were brick, lime, plastering material, Reinforced Cement Concrete (RCC) and steel generated from walls and foundations, column, lintel, roofing etc., totalling 270m^3 and measured the percentage of each of the waste materials generated. A mathematical model is developed such that the contractor can estimate the maximum revenue from demolition waste per m^3.	Ghosh et al. (2016)
This study deals with the sources, C&DW recycling plants, products manufactured and sold in India, with a case study in Delhi. It concludes that C&DW recycling is a vital part of reducing the environmental impact of the construction industry. The energy used and CO_2 emission from recycling processes must be balanced against any savings.	Ghosh and Ghosh (2016a, b)
This study presents a C&DW management framework to maximise the 3Rs, minimise the disposal of C&DW and consider a sustainable strategy using LCA (life cycle assessment). It develops a composite C&DW sustainability index for decisions on selection of material, sorting, recycle/reuse and treatment or disposal options.	Yeheyis et al. (2013)
This paper evaluates the impacts of two alternatives—recycling and disposing of C&DW and concluded that recycling reduces emission, energy use, and global warming potential (GWP) significantly, and conserves landfills space when compared to disposal of waste.	Marzouk and Azab (2013)
This article studies post-disaster C&DW observing that the increasing number of natural disasters has made post-disaster C&DW management a crucial component of disaster recovery in New Zealand after the Canterbury region suffered enormously from the 2010 and 2011 earthquakes. It estimated the generated post-disaster debris as 4 million tons and probably more than a million tons from repairs.	Domingo et al. (2011)
This study discusses the C&DW reuse, recycling and landfill status. Huge demand exists for aggregates in the housing and road sectors with a significant gap between demand and supply. Projections for building material requirement by the housing sector indicate a shortage of about 55,000 million m^3 aggregates and additional 750 million m^3 would be required to achieve the targets of the road sector. C&DW legislation is inevitably required with enhanced awareness level. Recycling technologies for C&DW need to be strengthened. The Bureau of Indian Standards should develop quality standards for the C&DW recycled products.	Ghosh et al. (2011)
This paper reports the feasibility of using crushed clay brick aggregates with less than 25% in weight, with fine clay brick aggregate with a range between 50–75%, as coarse derived from of C&DW generated from earthquakes as coarse and fine aggregates. These are used in producing non-structural partition wall blocks.	Xiao et al. (2011)
This study deals with the waste management strategies and their application in developing countries related to post-disaster through literature review and field survey. The results show that the strategies, issues and challenges vary according to type of disaster, magnitude, location, country, etc. The study identifies the issues and challenges of C&DW management.	Gayani et al. (2009)
This research presents the characteristics of demolition wastes, inert materials, coarse portions, and fines generated from earthquakes stored at three locations, namely, Puli, Tali and Taichung in Taiwan. The results shows that the volumetric content of inert materials from seismic demolition waste is about 96%, which can be reused as subbase and base soils due to its significant subgrade rating. The coarse aggregate also complies with Standard Taiwan Code CNS-490. The fines of inert materials show higher shear strengths than natural soils which can be reused as construction soils.	Yang (2009)

TABLE 1.2

Statistics of C&DW Generation, Recycling and Disposal

Country/ Region & Year	C&DW Generated	C&DW Debris Directed to Next Use	C&DW vs MSW Generated	C&DW Sent to Landfills	Composition of Demolition Waste in C&DW	Source
China 2020	2 billion tonnes	0.04 billion tonnes	80–90% of the total municipal waste	98%	97% (in 2013)	Qiao et al. (2020)
Dhaka, Bangladesh in 2018	1.28 million tonnes	1,088 – 1,152 tonnes	-	85–90% of generated C&DW	89% (1.139 million tonnes)	Islam et al. (2019), Islam (2016), Abedin and Jahiruddin (2015)
India 2020	150 million tonnes	-	1:6.5 (as approx. 960 MSW is produced/year)	95% (recycling rate is around 5%)	85%	
Iran 2017	1.2 million tonnes	-	50% of total MSW	75%		Asgari et al. (2017)
Spain 2011	47 million tonnes	-	25–30% of the total solid waste			Calvo et al. (2014), Coronado et al. (2011)
Sri Lanka 2019		3.1%		nearly 10–30%		Liyanage et al. (2019)
USA in 2018	600 million tonnes	600 million tonnes	C&DW two times that of MSW generated	24.2% (145 million tonnes)	90%	EPA: 2018 Fact Sheet ,USA, December 2020

on 12 May 2015, a Haiti earthquake in 2010, Hurricane Katrina in 2005, and an Indian Ocean tsunami in 2004 all generated volumes of waste which left the existing systems of waste management overwhelmed (Ranjitkar and Upadhyay 2015, Ulubeyl et al.,2017). Disaster debris in most of the cases impede rescuers and emergency services from reaching survivors. Disaster debris, namely, natural aggregates, C&D debris (concrete, bricks, timber, metal, etc), vehicles and boats, electrical goods and appliances can all be recycled (World Bank & United Nations, 2010). These recycled materials have a huge market in developing countries. Typical reuse of these materials include landfill cover, aggregate for concrete, fill for land reclamation and compost for fertilisation and slope stabilisation (Channell et al., 2009). Some materials can also be used beneficially to generate energy (Yepsen, 2008, USEPA, 2008).

Table 1.2 demonstrates the data of the amount of C&D debris generation, C&D debris going to next use for reuse and recycling, and the C&D debris sent to landfills, in various different countries and regions. In most of the countries, aggregate and brick are the main next use for the materials of the C&D debris.

1.3 Management and Recycling

A Waste Management plan is the significant part of a C&DW management strategy. (. Quantification of C&DW based on waste generation rates (WGR) is very important. The investigation of WGR has long been attractive to researchers and construction practitioners. In India there are only three major plants in Delhi and Ahmedabad for C&DW recycling and a few smaller plants in other locations. The recycling capacity in India is 6,500 tons per day (TPD) which is about 1% of the total C&DW generated. Recycled aggregates obtained through C&DW are of inferior quality and use of different pozzolanic materials are recommended by several researchers to enhance its properties. Fly ash is such a pozzolanic material that in the presence of water and calcium hydroxide produces cementitious compounds. Fly ash is a filler in hot mix asphalt applications and improves the fluidity of flowable fill and grout because of its spherical shape and particle size distribution. Effectiveness of C&DW recycling depends on the management plan of the C&DW that includes, storage, segregation & collection, recycled product design, pre-processing, if

TABLE 1.3

Debris Stream, Composition of C&DW, Potential Waste Recirculation, Landfilling and Business Models

Debris Stream, Composition, Type of Materials	Potential Waste Recirculation: Reuse/Recycle/Recover	Final Disposal	Business Model & Employment Generation
Construction and demolition, Building materials	Reuse, Recover & Recycle	Inert in sanitary landfill	Many examples already exist- C&DW Recycling business
Metals and white goods	Recover & Recycle	Inert in sanitary landfill	C&DW Recycling business model exists
Hazardous waste: Fuels, oil, batteries	Oil & batteries may be recycled	Inert in sanitary landfill	Business of Landfill
Municipal solid waste: Garbage, personal belongings	Recover & Recycle	Inert in sanitary landfill	Many examples already exist
Putrescible: Food waste, organic matter	Recycle		Business Models Exists
White goods, AC & vehicles; Refrigerators,	Reuse, Recover & Recycle	sanitary landfill	Business Models Exists
Special waste; Archaeological importance	Reuse, Recover & Recycle	sanitary landfill	Business Models Exists
Automobiles	Reuse, Recover & Recycle	Inert in sanitary landfill	Business Models Exists
Electronic waste	Reuse, Recover & Recycle	Inert in sanitary landfill	Business Models Exists
Inert Environmental Debris	-	sanitary landfill	
Treated wood	Reuse, Recover & Recycle		Business Models Exists

required, recycling, product quality compliance, storage, packaging and marketing. The business potential of C&DW is huge and the implementation of the business models using C&DW in increasing day by day.

The management of debris, which includes evacuation, separation, utilisation, and final disposal including other rehabilitation programmes is a major issue in the immediate and long-term recovery from such catastrophic events. The research experience shows the possibilities of developing reuse and recycling systems leading to effective business model and employment generation using the generated C&DW. Accordingly, Table 1.3 has been developed to gives glimpses on the potential of waste recirculation (Reuse/Recycling/Recovery), Final disposal to landfills, and business development. Literature concludes that the recycled construction materials produced using the C&DW generated from earthquakes can be used successfully as a replacement of virgin construction materials, compromising the quality to some extent in construction industries but that helps in reducing the consumption of natural resources and promoting the implementation of circular economy.

1.4 C&DW Legislation

There are many countries in the world where legislation on construction and demolition waste is not available or, even if it exists, it is not very robust. Hence the formal C&DW management has not been popular in most of the countries. However, informally the C&DW are managed and recycled in many developed, and developing even in least developed countries. C&DW recycling is increasingly becoming popular due its cost effectiveness and development of new technologies for the recycling. The other demotivating part is the poor awareness on use of recycled C&D products. Green building codes encourage the use of recycled C&D products in the building to a certain percentage.

US Environmental Protection Agency (USEPA) promotes a sustainable materials management (SMM) approach in C&D management. This will help in the reduction of natural resource extraction. In Austria, a Recycled Construction Materials Regulation was finally published on 29 June 2015, came into force on 1 January 2016 and lays down specific requirements to be complied with during construction or demolition of structures (Building Materials and Technology, 2015). This strengthened the pollutant investigation— mainly the suspended particulate matters, recycling-oriented demolition of structures and segregation of generated C&DW. The impact of C&DW on environment from the toxic dust particles from the debris is enormous. As per the ongoing National Clean Air Programme, cities have to reduce the SPM/RPM (Suspended Particulate Matters/Respiratory Particulate Matters) from the C&D debris, those are polluting the air by 20–30%, by 2024 (Sundaray & Bhardwaj, 2019). In India, the Construction and Demolition Waste management Rules were established in April 2016. As per legislation, all the cities must have a separate collection, storage and recycling system for the implementation of C&DW. In some of the cities in India, C&DW recycling plants have been installed and in many other cities are planning to have them in near future.

1.5 Discussion, Analysis, and Conclusion

It can be observed from the literature that there are huge gap exists in C&DW management from the desired situation. The governments and the entrepreneurs have to play a major role for resolving the issues and challenges. Following are a few points which may be considered for effective design of C&DW management systems for resource recovery, reduction of GHG emission and implementing the circular economy.

Robust system of quantification & estimation,	Availability of C&DW & tracking,
System of characterisation of C&DW	Financing for business case
Design of demolition technology for demolition of building and other infrastructure to gain mode C&DW in good condition	Design of standard supply chain for storage, segregation and collection and material recovery, Government involvement and support
Government involvement and support	Industry response and entrepreneurship
Standard operating procedure (SOP) for storage, segregation, collection and material recovery	Encouragement of utilisation of recycled C&DW in appropriate use
Availability of land for storage, collection, segregation, and recycling	Awareness on applicable C&DW management rules among the stakeholders across the nation
Quality Standard by national standard institution	Employment generation and livelihood earning
Standard quality and pricing of recycled C&DW products	Training on C&DW recycling technologies
Close monitoring of the system by local government and data management	Environmental compliance and dust management
Developing composite materials using C&DW	Research on C&DW management & recycling

There must be a pentagonal cooperation for the implementation of C&DW management systems in the country with the partnership of government, industries, researchers, (Non-Governmental Organisations) NGOs, and civil society. Effective C&DW management will definitely be helpful to reduce the materials consumption and production supporting the sustainable development goals.

ACKNOWLEDGEMENT

The author acknowledges the support of International Society of Waste Management, Air and Water (ISWMAW) and the Consortium of Researchers in International Collaboration (CRIC). The author acknowledges the support and guidance of Prof. Mija H. Hubler, Assistant Professor, Dept. of Civil, Environmental, and Architectural Engineering University of Colorado, Boulder, USA and Prof. Sadhan Kumar Ghosh, Professor of Dept. of Mechanical Engineering, and Ex-Dean, Faculty of Engineering and Technology, Jadavpur University, Kolkata, India.

REFERENCES

2020 Statewide Solid Waste Management Plan, May 2020. Moving toward zero waste, Delaware Solid Waste Authority, USA.

Abedin, M.A., and Jahiruddin, M., 2015. Waste generation and management in Bangladesh: An overview. *Asian Journal of Medical and Biological Research*, 1 (1), 114–120.

Akhtar, A., and Sarmah, A.K., 2018. C&D waste generation and properties of recycled aggregate concrete: a global perspective. *Journal of Cleaner Production,* 186, 262–281.

Alcántara-Ayala, I., 2002. Geomorphology, natural hazards, vulnerability and prevention of natural disasters in developing countries. *Geomorphology*, 47, (2–4), 107–124.

Aminu Lawan, A., and Angela, L.E.E. 2020. Is the concept of waste universal? Handling building demolition byproducts in the city of kano, Nigeria.

Asgari, A., Ghorbanian, T., Yousefi, N., Dadashzadeh, D., Khalili, F., Bagheri, A., Raei, M. and Mahvi, A.H. 2017. Quality and quantity of construction and demolition waste in Tehran. *Journal of Environmental Health Science & Engineering.* doi 10.1186/s40201-017-0276-0

Brandão, R., Edwards, D.J., Hosseini, M.R., Silva Melo, A.C. and Macêdo, A.N. 2021. Reverse supply chain conceptual model for construction and demolition waste. https://doi.org/10.1177/0734242X21998730

Building Materials and Technology Promotion Council. 2018. Utilisation of recycled produce of construction & demolition waste: a ready reckoner.

Calvo, N., Varela-Candamio, L. and Novo-Corti, I. 2014. A dynamic model for construction and demolition (c&d) waste management in Spain: driving policies based on economic incentives and tax penalties; *Sustainability* 2014, 6, 416–435; doi:10.3390/su6010416

Channell, M., Graves, M. R., Medina, V. F., Morrow, A. B., Brandon, D. and Nestler, C. C. 2009. Enhanced tools and techniques to support debris management in disaster response missions. US Army Corps of Engineers. Vicksburg, MS: Environmental Laboratory US Army Engineer Research and Development Center.

Construction and Demolition Waste Management in Austria, V2 – September 2015; © 2014 Deloitte SA. Member of Deloitte Touche Tohmatsu Limited

Coronado, M., Dosal, E., Coz, A., Viguri, J. R., Andre, A.2011. Estimation of construction and demolition waste (C&DW) generation and multicriteria analysis of C&DW management alternatives: A case study in Spain, *Waste and Biomass Valorization* 2: 209–225. doi 10.1007/s12649-011-9064-8

Domingo, N., Luo, H.J. and Egbelakin, T. 2011. Composition of demolition wastes from Chi-Chi earthquake-damaged structures and the properties of their inert materials.

Gálvez-Martos, J.L., Istrate, I.R., 2020. Construction and demolition waste Management; DOI: 10.1016/B978-0-12-819055-5.00004-8

Ghaffar, S.H., Burman, M., and Braimah, N. 2020. Pathways to circular construction: An integrated management of construction and demolition waste for resource recovery, *Journal of Cleaner Production*, 244, 118710, DOI: 10.1016/j.jclepro.2019.118710.

Ghosh, S. and Ghosh, S. 2011. Rebuilding C&D Waste Recycling Efforts in India, *Waste Management World*, September 2011 issue, Volume 12, Issue 5.

Ghosh, S. and Ghosh, S. 2016a. Construction and Demolition Waste. Sustainable Solid Waste Management, book edited by J.W.C. Wong; Rao Y. Surampalli; T.C. Zhang; R.D. Tyagi; and A. Selvam, Published by the *American Society of Civil Engineers* (pp. 511–547), DOI: 10.1061/9780784414101.ch16.

Ghosh, S. and Ghosh, S.. 2016b. Construction and Demolition Waste. 10.1061/9780784414101.ch16.

Ghosh, S.K., Haldar, H.S., Chatterjee, S. and Ghosh, P. 2016. An optimization model on construction and demolition waste quantification from building, *Procedia Environmental Sciences* 35 (2016), 279–288.

Guha-Sapir, D., Vos, F., Below, R., Ponserre, S. Annual disaster statistical review 2011: the numbers and trends. CRED, Brussels, 2012. Disponívelem: <http://www.cred.be/sites/default/files/ADSR_2011.pdf>.

Gayani Karunasena, Sajani Jayasuriya, 2013. Construction safety assessment framework for developing countries: a case study of Sri Lanka, *Journal of Construction in Developing Countries* 18(1):33–51

ICIMOD. http://www.icimod.org/v2/cms4/_files/images/e92e3b0202d11e51262a6e2cb1ed6f2d.jpg

IFRC World Disasters Report 2010. Focus on urban risk. International Federation of Red Cross and Red Crescent Societies (IFRC), Geneva, 2010. Available at: http://www.ifrc.org/Global/Publications/disasters/WDR/wdr2010/WDR2010-full.pdf

Islam, F.S. 2016. Solid waste management system in Dhaka City of Bangladesh. *Journal of Modern Science Technology* 4 (1), 192–209

Islam, R., Nazifa, T.H., Yuniarto, A., Uddin, A.S., Salmiati, S. and Shahid, S., 2019. An empirical study of construction and demolition waste generation and implication of recycling, *Waste Management*, 95 (2019) 10–21.

Jain, M.S. 2021. A mini review on generation, handling, and initiatives to tackle construction and demolition waste in India; https://doi.org/10.1016/j.eti.2021.101490.

Jin, R., Li, B., Zhou, T., Wanatowski, D., Piroozfar, P. 2017. An empirical study of perceptions towards construction and demolition waste recycling and reuse in China. *Resources, Conservation and Recycling* 126, 86–98.

Karunasena, G., Amaratunga, D., Haigh, R. & Lill, I. 2009. Post disaster waste management strategies in developing countries: Case of Sri Lanka.

Liyanage, K.L.A.K.T., Waidyasekara, K.G.A.S., Mallawaarachchi, B.H. and Pandithawatta, T.P.W.S.I. 2019. Origins of Construction and Demolition Waste Generation in the Sri Lankan Construction Industry; Proceedings of the World Conference on Waste Management, Vol. 1, pp. 1–8; DOI: https://doi.org/10.17501/26510251.2019.1101

Marzouk, M., Azab, S. 2013. Environmental and economic impact assessment of construction and demolition waste disposal using system dynamics. http://dx.doi.org/10.1016/j.resconrec.2013.10.015.

Poon, C.S., Yu, A.T., Wong, A., Yip, R., 2013. Quantifying the impact of construction waste charging scheme on construction waste management in Hong Kong.

Qiao, L., Liu, D, Yuan, X, Wang, Q. and Ma, Q. 2020); generation and prediction of construction and demolition waste using exponential smoothing method: a case study of Shandong Province, China; *Sustainability*, 12, 5094; doi:10.3390/su12125094

Ranjitkar, M.G. and Upadhyay, S.. 2015. Post-earthquake debris management: challenges and opportunities in Nepal, *A Journal of Rural Infrastructure Development* 6 6, 2015.

Sundaray, N.K. and Bhardwaj, S.R., Ed. 2019. NCAP-National Clean Air Programme (NCAP), Ministry of Environment, Forest & Climate Change, Government of India,

Ulubeyl, S., Kazazb, A., Arslana, V. 2017. Construction and demolition waste recycling plants revisited: management issues. DOI: 10.1016/j.proeng.2017.02.139

USEPA 2008. Planning for natural disaster debris. Washington: Office of Solid Waste and Emergency Response, Office of Solid Waste, USEPA.

Vidyasekar, K.G.S.B. 2019. Implementation of 3R principle in construction and demolition waste management *International Journal of Innovative Technology and Exploring Engineering*, 8, 12, pp. 977–980, DOI: 10.35940/ijitee.

World Bank & United Nations. 2010. Natural hazards, unnatural disasters: the economics of effective prevention. Washington, DC: The International Bank for Reconstruction and Development/The World Bank,.

Xiao, Z., Ling, T.C., Kou, S.C., Wang, Q., Poon, C.S. 2011. Use of wastes derived from earthquakes for the production of concrete masonry partition wall blocks.

Yang, C. P. 2009. Composition of demolition wastes from Chi-Chi earthquake-damaged structures and the properties of their inert materials. *Construction Engineering Management* 139 (5), 466–479.

Yeheyis, M., Kasun, H., Shahria Alam, M., Eskicioglu, C, Sadiq, R 2013. An overview of construction and demolition waste management in Canada: a lifecycle analysis approach to sustainability. DOI : 10.1007/s10098-012-0481-6

Yepsen, R. 2008. Generating Biomass fuel from disaster debris. *Biocycle,* 49(7) 51.

Zhao, W., Leeftink, R.B., Rotter, V.S., 2010. Evaluation of the economic feasibility for the recycling of construction and demolition waste in China—the case of Chongqing. *Resources, Conservation and Recycling* 54 (6), 377–389.

2

Circular Economy in Construction: An Overview with Examples from Materials Research

Dhanada K Mishra
KMBB College of Engineering and Technology (BPUT), Odisha, India

Jing Yu
Sun Yat-Sen University, Guangzhou, PR China

CONTENTS

2.1 Background 11
2.2 Introduction 13
2.3 Circular Economy and Construction Industry 14
2.4 Examples—Novel Construction Materials 14
    2.4.1 Limestone-Calcined Clay (LCC) Pozzolana and Cement 15
    2.4.2 'Smart' Composites 17
    2.4.3 Recycled PET Fibre Reinforced Cementitious Composite 18
2.5 Summary and Discussion 19
2.6 Conclusions and Recommendations 19
Acknowledgments 20
Notes 20
References 20

2.1 Background

The idea of Circular Economy (CE) goes back to the 1960s when the concept of the open economy, as opposed to the closed economy, was discussed by Boulding (1966). The 'use and throw' model of the 'Linear Economy' was replaced by the 'Make, Use and Re-use' model (Figure 2.1) for the first time in the report by Walter Stahel and Genevieve Reday for European Commission in 1976 (Stahel & Reday, 1981). The Organisation for Economic Co-operation and Development (OECD) defines a circular economy as a system which maximises the value of the materials and products that circulate within the economy (OECD, 2020). Allowing for sharp resources preservation and environmental footprint reduction, circular business models are attracting a growing attention from researchers, governments, and industries. The best way to understand 'Circular Economy' is by examining a simple real-life case study such as lighting, for instance. It has always been known that a light bulb could be designed to last a lifetime, but it was never profitable to do so for obvious reasons. However, the current trend is to lease lighting as a service rather than lights as a product which is phrased by the lighting company Phillips as—'Circular Lighting'[1]. Such an approach vastly reduces the huge quantity of discarded light bulbs from going to landfill. It incentivises the manufacturer to build the product to last as long as possible, design it for easy recycling and also take responsibility for the entire lifecycle of the product. The iconic manufacturer of electronic products, *Apple,* is another good example. The company announced in 2017[2] that they will be

DOI: 10.1201/9781003217619-3

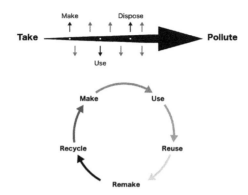

FIGURE 2.1 Schematic depiction of circular economy (right in colour) compared to the linear economic models (left in grey).(Weetman, 2016).

making all their new iPhones, iMacs, and other products from 100% recycled materials. The company has been well known for collecting back all its products at the time of an upgrade. It has a substantial R&D programme to make the transition to circular product manufacturing, including in the use of paper and energy. However, it is sobering that 53.6 million tonnes of e-waste were generated globally in 2019 out of which only 17.4% were properly documented and recycled (Forti et al., 2020). Although growing at an increasing rate, the recycling effort is not able to keep pace with the growth in generation of e-waste.

Modern lifestyle dependent on the linear industrial processes, uses finite natural resources to create products with a limited-service life, that finally end up in landfills or in incinerators. On the other hand, CE is inspired by living systems like organisms that process nutrients that can eventually be fed back into the production cycle. Biomimicry, 'cradle to cradle' instead of 'cradle to grave', closed loop, or regenerative processes are some of the other terms usually associated with it. The World Economic Forum (WEF), World Resources Institute (WRI), Philips, Ellen MacArthur Foundation, UN Environment Programme (UNEP), and 40 other partners launched the Platform for Accelerating the Circular Economy (PACE)[3] in 2018 to scale up circular economy innovations. PACE had three focus areas, namely, blended finance (especially for developing countries), policy frameworks, and public-private partnerships. Global corporations like IKEA, Coca-Cola, Alphabet Inc., and DSM (company), along with governments of Denmark, The Netherlands, Finland, Rwanda, UAE, China, etc. are members of PACE. China's 11th five-year plan included the promotion of a circular economy as a national policy beginning in the year 2006. The British Standards Institution (BSI) developed and published the first standard for CE—"BS 8001:2017 Framework for implementing the principles of the circular economy in organizations"—in 2016. An article by McKinsey titled *"Remaking the industrial economy"* has identified the potential for significant benefits with global economic impact from net materials cost savings worth up to $1 trillion annually by 2025. The four areas of benefit from reconfiguration of key manufacturing processes and flows of materials and products identified are—net savings of materials, mitigated supply risks, innovation potential, and job creation. Most importantly, CE can contribute to meeting the emission reduction goals of the COP 21 Paris Agreement. Since, the greenhouse gas (GHG) reduction commitments made by signatory countries, are not sufficient to limit global warming to 1.5°C, it is estimated that half of the additional emissions reductions of 15 billion tonnes CO_2 per year needed can be delivered by CE (Blok et al., 2017).

In 2012, Kate Raworth, a senior research associate at Oxford University introduced the idea of a Doughnut Economy (see Figure 2.2) that aims to make human welfare the basis of economic policy rather than the all-pervading pursuit of Gross Domestic Product (GDP) growth (Raworth, 2012). Economic welfare was defined as per the Sustainable Development Goals (SDGs) that provide a set of minimum living standards by the UN for every human being. She extended the basic circular flow of money and goods model of economics by adding the nine planetary boundaries (environmental limits imposed by the planet) on the outside and social boundary consisting of SDGs on the inside—forming a doughnut. In this model, economic progress is measured in terms of the balance between human wellbeing and protection of the life-support systems provided by the planet, thus mitigating global warming, ecological breakdown and climate change. Since its introduction, this post-growth economic thinking has attracted the attention

Circular Economy in Construction

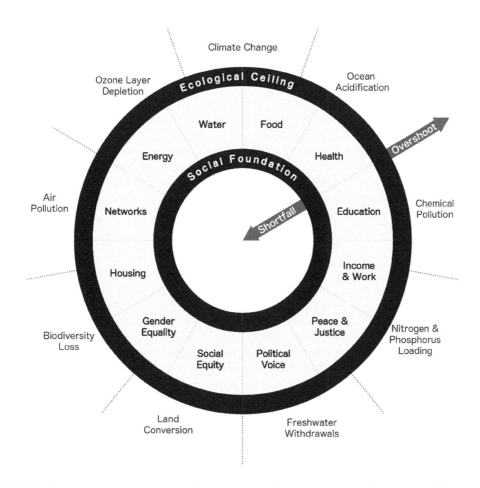

FIGURE 2.2 Schematic depiction of the idea of Doughnut Economic model proposed by Kate Raworth (2012).

of a range of actors ranging from the UN General Assembly to Occupy the London movement. As many countries start re-examining prevalent strategies and economic policies for a green recovery in light of the current Covid-19 pandemic, the foundations of the Doughnut Model are being explored for the long-term planning and policy development of cities like Amsterdam.

2.2 Introduction

The Intergovernmental Panel on Climate Change (IPCC) recommendations require a substantial reduction in emission of Green House Gases (GHGs) to a net zero level by 2050 in order to limit global warming to below 1.5°C above pre-industrial global mean surface temperatures (IPCC, 2018). The fast-depleting carbon budget to achieve this target has triggered a global effort to find low carbon alternatives to existing technologies in every carbon intensive sector such as energy, transport and building construction. The contribution of the building construction sector to the global emission is estimated to be anywhere between 39% of the total, which also accounts for 36% of final energy use (UNE & IEA, 2017). The cement production alone accounts for 6–7% of the same (Boden et al., 2017). Portland cement, the primary manufactured component of concrete, contributes almost one kg of CO_2 per kg of cement produced from the decomposition of calcium carbonate and use of fossil fuel (Monteiro et al., 2017). There is a global effort to address this challenge in four possible ways, e.g., reducing cement content in concrete, reducing clinker content in cement, reducing energy consumption in manufacturing of cement by switching to low carbon fuels, and improving thermal and electrical efficiencies. Global cement production is projected to increase on average 12–23% by 2050. This would result in a 4% increase in CO_2 emissions globally in 2050 under the International Energy Agency (IEA) Reference Technology Scenario (RTS), despite a 12% increase in global cement production during the same time frame. The April 2018 revision of the Cement Industry Technology Roadmap, developed by the IEA and Cement Sustainability Initiative (CSI) takes into account the December 2015 Paris Climate Agreement's objective to limit the rise in global temperatures this century to less than 2°C above pre-industrial levels. Realising this 2-degree Celsius Scenario (2DS) by 2050 implies a 24% reduction in current levels of global direct CO_2 emissions from cement manufacture, despite a projected increase in global cement production. The roadmap vision requires 7.7 Gigatonnes CO_2 cumulative direct carbon emissions savings from cement making by 2050, compared to the RTS. This can only be achieved by adopting all the four strategies mentioned above in addition to innovative future technologies such as carbon capture and storage. The framework of circular economy is going to play an important role as we discuss in the following sections using three examples from the innovations in construction materials technology.

2.3 Circular Economy and Construction Industry

Besides being a large contributor to greenhouse gas emissions, the construction sector is also one of the world's largest waste generators. The waste generated during construction and demolition of old structures grew at about 5% annually and averaged around 3 billion tonnes globally in 2012 (Akhtar & Sarmah, 2018). This waste consisted of materials such as bricks, metals, plastics, wood & timber, gypsum, stones, sanitary ware, and concrete. The circular economy approach is an effective solution to diminish the environmental impact of the industry by application of the '3Rs' or Reuse/Recover/Recycle principles.

The World Green Build Council in its 2019 report declared a vision that sets the following ambitious goals (WGBC, 2019) that may have far reaching impact.

- By 2030, all new buildings, infrastructure and renovations will have at least 40% less embodied carbon with significant upfront carbon reduction, and all new buildings will have net zero operational carbon.
- By 2050, new buildings, infrastructure and renovations will have net zero embodied carbon, and all buildings, including existing buildings must be net zero operational carbon.

The four strategies to achieve the above are prevent, reduce, plan, and offset. They involve prevention of new construction by better utilisation of existing infrastructure, reduction in use of carbon intensive materials in favour of low carbon alternatives, planning for future use and end of life disposal and also use carbon offset strategies to achieve net zero carbon goals.

Decision-making about the circular economy can be performed on the operational (connected with the specific portions of the production process), tactical (connected with the whole process), and strategic (connected with the whole organisation) levels. It may concern both construction companies as well as construction projects (where a construction company is one of the stakeholders). A first attempt to measure the success of the circular economy implementation was done in a construction company (Nuñez-Cacho et al., 2018, Górecki et al., 2019). The circular economy can contribute to creating new jobs and economic growth. One of such posts, for example, may be the circular economy manager employed for construction projects. In addition, the opportunities for new job creation in the recycling industry is enormous.

2.4 Examples—Novel Construction Materials

In 2019, the level of cement consumption worldwide is estimated to have reached 4,100 million tonnes (USGS, 2020), an increase of 60% since 2006 (Cemnet, 2017). The top 20 countries represent 85% of global cement consumption. China alone, consumes 2,347 million tonnes, representing 57%. Together

Circular Economy in Construction

with India, at 297 million tonnes and as the second largest producer, these two countries represent 64% of total global cement consumption. There is a global effort to produce greener cement and concrete by using alternative binders and Cement Replacement Materials (CRM) (Yu et al., 2017, Yu et al., 2018a, b). It is likely the amount of usable CRM such as fly ash and ground granulated blast furnace slag available globally (around 10% of the Portland Cement production) is inadequate to satisfy the demand from the construction industry (Scrivener, 2014). In this context, Calcined Clay is rapidly emerging as an alternative material that is widely and adequately available. In recent years, Limestone-Calcined Clay (LCC) has been proposed as a new type of green CRM to help continue the trend of reducing clinker content in blended cement and resulting concrete (Lothenbach et al., 2011, Zunino & Scrivener, 2020, Dhandapani et al., 2018). In the following examples, we discuss the recent research results exploring use of ultra-high-volume replacement of Ordinary Portland Cement (OPC) with LCC as a pozzolan. The second example we discuss, is about the application of self-sensing and self-healing smart composites that can extend the service life of concrete infrastructure resulting in net savings of raw materials. The final example talks about the use of recycled Polyethylene Terephthalate (PET) fibre as a replacement of much more expensive virgin polyvinyl alcohol (PVA) fibre in cement-based composites and its environmental and cost implications.

2.4.1 Limestone-Calcined Clay (LCC) Pozzolana and Cement

Use of high volume pozzolanic substitute for Portland Cement has been attempted for many decades to achieve lower heat of hydration, reduction in cost and improved sustainability. Before the advent of pozzolans like fly ash and ground granulated slag, calcined clay was used in many large infrastructure projects such as dams and foundations that involved mass concreting. Such practice was discontinued with the availability of cheap industrial waste products like fly ash and slag as coal-fired thermal power plants cropped up everywhere as the primary source of electric power. However, recent concerns about greenhouse gas emissions causing global warming leading to climate crisis has resulted in a shift away from thermal power and hence use of fly ash is under decline. In any case the amount and availability of quality fly ash and slag is not adequate for the projected global demand of the construction sector. Limestone-Calcined Clay (LCC) is recently emerging as a promising pozzolan along with the Limestone-Calcined Clay Cement (LC3). Figure 2.3 shows the composition of this green cement which consists of 45% clinker, 5% gypsum, 35% calcined clay and 15% limestone powder. This cement performs as well as OPC as shown in Figure 2.4. Except for early strength at 1-day and 3-days, the LC3 out-performs OPC and PPC (Pozzolana Portland Cement) at all other ages. In some respects, such as durability may even out-perform OPC.

The use of LCC as a pozzolan is also gaining ground since it has certain advantages in being more widely available and having a higher quality assurance being a manufactured product. Recently research has been carried out to explore the possibility of using LCC at a binder replacement level higher than 60%. Figure 2.5 shows the compressive strength development of a range of mortars with LCC replacement

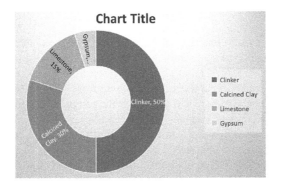

FIGURE 2.3 Composition of Limestone-Calcined Clay Cement (LC3).

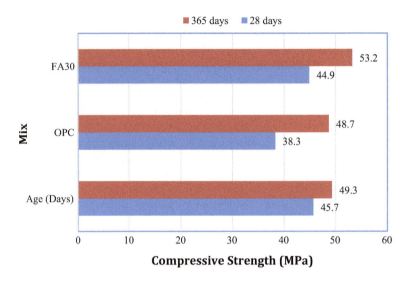

FIGURE 2.4 Composition of compressive strength of Limestone-Calcined Clay Cement (LC3) with OPC, PPC and PSC. (Dhandapani et al., 2018.)

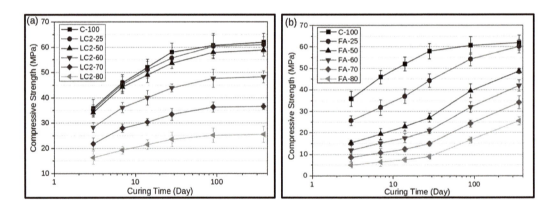

FIGURE 2.5 Compressive strength of standard mortar with SCM at different replacement levels from 3–360 days: (a) LC2-Series; (b) FA-Series. (Yu et al., 2018a, b.)

levels from 25–80%. When used in excess of 60% as a fraction of the binder, high-volume pozzolanic concrete can reduce compressive strength significantly at an early age. In the current study, experimental results for concrete samples are reported that capture the effect of LCC dosage at five different percentages (25%, 50%, 60%, 70% and 80%) at testing age ranging from 3 days to 365 days. It is observed that with judicious choice of water binder ratio and SP dosage, the compressive strength of mortar or concrete can reach over 35 and 45 MPa respectively at 7-day age, and over 43 MPa and 56 MPa at 28-day age, even when 70% of binder material is replaced by LCC. The results were compared to a commercial 100% cement mortar and Grade 45 concrete. Around 48% reduction in CO_2 emission and more than 20% reduction in embodied energy observed in the ultra-high volume LCC concrete. Figure 2.5 shows corresponding results for ultra-high-volume fly ash mortars that show similar trends except that the rate of strength development is conspicuously slower (Table 2.1).

TABLE 2.1
Major Countries in Worldwide Cement Production from 2015—2019 (in Million Metric Tonnes)

	2015	2016	2017	2018	2019*
China	2,350	2,410	2,320	2,200	2,200
India	270	290	290	300	320
Vietnam	61	70	78.8	90.2	95
USA	83.4	85.9	86.6	87	89
Egypt	55	55	53	81.2	76
Indonesia	65	63	65	75.2	74
Iran	65	53	54	58	60
Russia	69	56	54.7	53.7	57
Brazil	72	60	53	53	55
South Korea	63	55	56.5	57.5	55
Japan	55	56	55.2	55.3	54
Turkey	77	77	80.6	72.5	51
Rest of the world	473	524	706	866	914
Total	**3,758**	**3,855**	**3,953**	**4,050**	**4,100**

2.4.2 'Smart' Composites

One aspect that is getting increasing research attention of late is the inherent weakness of steel reinforced concrete in terms of its relatively short service life and the possibility of extending the same by use of 'Smart' Cement based composites. Depending on the importance of the structure (e.g., nuclear power plants), real-life structures are designed for a lifetime ranging between 50 to 100 years and occasionally up to 200 years or more. It is difficult to accurately model all the degradation processes for construction materials like concrete making up these structures in short duration experiments or field observations. It is also difficult to have sufficiently long-term field data that is useful for many materials (e.g., high-performance concrete) with the nature of the material itself constantly changing. Reinforced concrete is greatly affected by corrosion of the steel reinforcements and accelerated ageing tests may not accurately capture all the environmental processes that cause degradation. For different exposure conditions, design codes specify prescriptive design requirements such as limits on water/cement ratio, crack width, deflection, etc. Only recently codes such as the European fib 34 Model code for service life design have started addressing explicit performance-based design approach to service life (fib Task Group 5.6, 2006). Smart cement-based composites that have self-sensing and self-healing properties (Figure 2.6) can help overcome the aforementioned challenges in a long-term and robust structural health monitoring scheme.

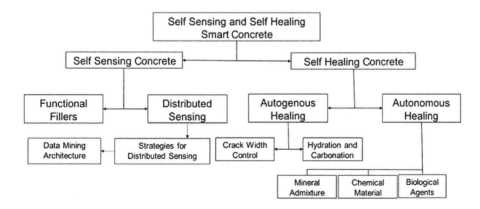

FIGURE 2.6 Classification of self-sensing and self-healing smart composites. (Das et al., 2019.)

It may even allow for continuous learning and prescribing appropriate repair and maintenance regimes (Das et al., 2019). Impact of such materials combined with the green technologies described above would result in a green and smart concrete that would make the future of civil engineering infrastructure truly sustainable.

2.4.3 Recycled PET Fibre Reinforced Cementitious Composite

The enormous quantities of Polyethylene Terephthalate (PET) solid wastes have led to serious environmental issues. They constitute an important portion of the total plastic waste but lag in reuse and recycling. Recent studies have explored the feasibility of recycling PET solid wastes as short fibres in Strain-Hardening Cementitious Composites (SHCCs), which exhibit strain-hardening and multiple cracking under tension, and therefore have clear advantages over conventional concrete (being prone to cracking being weak in tension) for many construction applications (Yu et al., 2018a, b). The hydrophobic PET surface was treated with NaOH solution followed by a silane coupling agent to achieve the dual purpose of improving the fibre/matrix interfacial frictional bond (from 0.64 MPa to 0.80 MPa) and enhancing the alkali resistance, which is important for applications in an alkaline cementitious environment. With surface treatment, recycling PET wastes as fibres in SHCCs is a promising approach to significantly reduce the material cost of SHCCs while disposing hazardous PET wastes in the construction industry. Figure 2.7 shows the tensile stress-strain behaviour of such SHCC mixes in which the more commonly used expensive PVA fibre has been replaced by treated PET fibre.

Table 2.2 summarises the Material Sustainability Indicators (MSIs) per unit volume of conventional Grade 45 concrete, a typical SHCC with a FA/binder ratio of 55% denoted as M45, and an ultra-high-volume fly ash with FA/binder ratio of 80% and PVA/PET fibre reinforced SHCCs (mixes P20 and U20 respectively with 2% fibre by volume). Due to the presence of polymeric fibres, SHCCs normally account for higher embedded energy than conventional concrete. By introducing ultra-high-volume fly ash in P20, it consumes about 30% less energy than the typical SHCC (M45). In terms of the indices of CO_2 emission and solid waste use, the material sustainability performance of P20 surpasses that of conventional concrete. In comparison to P20, U20 exhibits a decrease of around 37.4% in energy, 7.5% in carbon footprint and a reduction of about 78.6% in cost. This shows the great advantage of using the circular economy approach promoting reuse and recycle of industrial waste like fly ash and used PET.

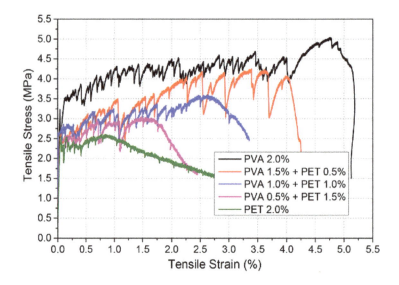

FIGURE 2.7 Tensile stress-strain relation of HyECC with 2 vol% PVA/recycled PET fibres. (Yu et al., 2018a, b)

TABLE 2.2

Material Sustainability Indicators and Cost Comparison for Different Mixtures (Yu et al., 2018a, b)

Sl. No	Mixture	Embodied Energy (GJ/m³)	CO_2 Emission (kg/m³)	Solid Waste (kg/m³)	Cost (HKD/m³)
1	Grade 45 Concrete	2.76	450	0	587
2	Typical SHCC (M45)	6.07	606	−691	4,883
3	Ultra-High-Volume Fly ash SHCC w/PVA fibre (P20)	4.17	293	−996	4,472
4	Ultra-High-Volume Fly ash SHCC w/PET (P20)	2.61	271	−1,024	955

2.5 Summary and Discussion

During his testimony to the US congress in 1973, Kenneth Boulding had said - 'Anyone who believes that exponential growth can go on forever in a finite world is either a madman or an economist'. In recent times, it has become increasingly clear that a free market-driven open-ended growth model is unsustainable and is the primary driver of global heating leading to the ecological breakdown and climate crisis. Today, the world is staring at a climate crisis proving the truth of his words. Circular economy models can be the basis for a progressive economic and business framework that can reverse this trend. While their applicability in sectors such as consumer products is straight-forward, the construction sector would present a more complicated set of challenges. The creation of positions such as circular economy manager in construction projects, restructuring of construction contracts making the contractor responsible for maintenance and end of life recycling/reuse of built infrastructure, use of industrial waste in cement and concrete, are all steps in this direction. An excellent example of such an initiative has been undertaken by the government of The Netherlands in the form of a circular infrastructure business model (Geradts, 2020). Treating viaducts and bridges as a service, this model requires modular construction with buyback guarantee at the end of the design service life. Funded by the government, infrastructure companies together with clients jointly operate the facility in PPP mode. The companies responsible for design, materials choice and construction will also be the ones connected to the long-term value of the building blocks and eventual dis-assembly and reuse. The sooner such developments are made mandatory by progressive codes and regulations, the better for the environment and future sustainability.

2.6 Conclusions and Recommendations

The new approach of circular economy comprising 'reduce, reuse and recycle' as applied to the construction sector has been explained with the help of three practical examples from recent research in the field of construction materials. On the basis of the discussions the following conclusions and recommendations can be made.

- In light of the impending climate crisis, the circular economy approach must be adopted in all sectors of economy to achieve the goal of 2015 Paris climate agreement and limit the increase of temperature to below 2°C. This would require that the construction industry must ensure significant reduction in its projected emission of greenhouse gases.
- Cement, being one of the major sources of greenhouse gas emission, must bear a major responsibility to achieve the above target by reducing clinker factor, energy consumption and switching to low carbon fuel besides using carbon capture and storage.
- The development of Limestone-Calcined Clay Cement (LC3) and potential for use of ultra-high-volume replacement of binder in concrete by Limestone-Calcined Clay as a pozzolan is a great example of use of circular economy-based approach towards meeting the IPCC's goals.

- Use of structural health monitoring (SHM) by deploying self-sensing material and extending life of concrete infrastructure by use of self-healing green composites is another strategy which can help achieve the goal of reducing carbon emission from the construction sector.
- The utilisation of recycled PET fibres in SHCCs significantly reduces the environmental impact and material cost. When half of the PVA fibres are replaced by recycled PET fibres, the composite shows a decrease of around 20% in energy intensity and a reduction of about 40% in cost.
- It is recommended that a circular economy-based approach must be made mandatory for use in the construction industry in order to drive innovation and help avoid climate crisis.

ACKNOWLEDGMENTS

Dhanada K Mishra gratefully acknowledges KMBB College of Engineering and Technology, Bhubaneswar, India for permitting his sabbatical and Hong Kong University of Science and Technology (HKUST) for providing the facilities where much of the research reported was carried out.

NOTES

1 "Philips delivers on commitment to the circular economy at DAVOS 2019". Philips. Retrieved 2020-07-10.
2 https://www.greenbiz.com/article/can-apple-close-loop-tech-giant-targets-100-recycled-material Accessed: 2020-07-10.
3 Hub, IISD's SDG Knowledge. "WEF Launches Public-Private Platform on Circular Economy | News | SDG Knowledge Hub | IISD". Retrieved 2020-07-10.

REFERENCES

Akhtar, A., and Sarmah, A.K. (2018). Construction and demolition waste generation and properties of recycled aggregate concrete: A global perspective. *Journal of Cleaner Production*, 186, 262–281.

Blok, K., Hoogzaad, J., Ramkumar, S., Ridley, S., Srivastav, P., Tan, I., Terlouw, W., and Terlouw, Wit de. 2017. Implementing circular economy globally makes Paris targets achievable. *Circle Economy*, Ecofys. Retrieved 20 April 2017.

Boden, T.A., Marland, G., and Andres, R.J. (2017). *Global, Regional, and National Fossil-Fuel CO$_2$ Emissions*, Carbon Dioxide Information Analysis Center, Oak Ridge National Laboratory, US Department of Energy, Oak Ridge, Tenn., USA

Boulding, K.E. (1966). The economics of the coming spaceship earth, In: H. Jarrett (ed.), *Environmental Quality Issues in a Growing Economy*. Baltimore, MD: Resources for the Future/Johns Hopkins University Press, pp. 3–14.

Cemnet. (2017). *Global Cement Report*. 12th edition. Cement review.

Das, A.K., Mishra, D.K., Yu, J., and Leung, C.K.Y. (2019). Self-sensing and self-healing 'smart' and green cement-based composites—a review from service life design perspective. Special issue on "High-performance cementitious materials" for ASTM's journal *Advances in Civil Engineering Materials* 8 (3). 554–578. doi:10.1520/ACEM20190023.

Dhandapani, Y., Sakthivel, T., Santhanam, M. Gettu, R., and Pillai, R. G., (2018). Mechanical properties and durability performance of concretes with Limestone Calcined Clay Cement (LC3). *Cement and Concrete Research*, 107, 136–151.

fib Task Group 5.6. (2006). Model code for service life design, Lausanne, Switzerland: fédération internationale du béton.

Forti V., Baldé C.P., Kuehr R., and Bel G. (2020). The Global E-waste Monitor 2020: Quantities, flows and the circular economy potential. United Nations University (UNU)/United Nations Institute for Training and Research (UNITAR)—co-hosted SCYCLE Programme, International Telecommunication Union (ITU) & International Solid Waste Association (ISWA), Bonn/Geneva/Rotterdam. 120 p.

Geradts, M. (2020). Circular infrastructure business models: Insights from the open learning environment circular bridges and viaducts, Platform for Accelerating the Circular Economy (PACE), 26 p.

Górecki, J., Nuñez-Cacho, P., Corpas-Iglesias, Francisco, A., Molina-Moreno, V., (2019). How to convince players in construction market? Strategies for effective implementation of circular economy in construction sector. *Cogent Engineering* 6 (1), 1–22. doi: 10.1080/23311916.2019.1690760

Intergovernmental Panel on Climate Change (IPCC). (2018). Global warming of 1.5°C. An IPCC Special Report on the impacts of global warming of 1.5°C above pre-industrial levels and related global greenhouse gas emission pathways, in the context of strengthening the global response to the threat of climate change, sustainable development, and efforts to eradicate poverty [Masson-Delmotte, V., P. Zhai, H.-O. Pörtner, D. Roberts, J. Skea, P.R. Shukla, A. Pirani, W. Moufouma-Okia, C. Péan, R. Pidcock, S. Connors, J.B.R. Matthews, Y. Chen, X. Zhou, M.I. Gomis, E. Lonnoy, T. Maycock, M. Tignor, and T. Waterfield (eds.)]

Nuñez-Cacho, P., Górecki, J., Molina-Moreno, V., Corpas-Iglesias, F.A. (2018). What gets measured, gets done: Development of a circular economy measurement scale for building industry. *Sustainability*. 10(7) 2340. doi: 10.3390/su10072340

Lothenbach, B., Scrivener, K., and Hooton, R.D. (2011). Supplementary cementitious materials. *Cement and Concrete Research*, 41, 1244–1256.

Monteiro, P.J.M. Miller, S.A. and Horvath, A. (2017). Towards sustainable concrete. *Nature Materials*, 16, 698–699.

OECD. (2020). Circular economy, waste and materials, *Environment at a Glance: Indicators*, http://www.oecd.org/environment-at-a-glance. (Accessed: 2020-07-12)

Raworth, K. (2012). A safe and just space for humanity, *Oxfam Discussion Paper*, p. 26.

Scrivener, K.L. (2014). Options for the future of cement. *The Indian Concrete Journal*, 88, 11–21.

Stahel, W. and Reday, G. (1981). *Jobs for Tomorrow, the Potential for Substituting Manpower for Energy*, Vantage Press, N.Y. 116.

UN Environment and International Energy Agency. (2017). Towards a zero-emission, efficient, and resilient buildings and construction sector. Global Status Report. 48 p.

US Geological Survey (USGS). (2020). Mineral commodity summaries, U.S. Geological Survey, Reston, Virginia. USA. 200 p., https://doi.org/10.3133/mcs2020

Weetman, C. (2016). A Circular Economy Handbook for Business and Supply Chains: Repair, Remake, Redesign, Rethink. London, United Kingdom: Kogan Page. p. 25. ISBN OCLC967729002.

WGBC. (2019). *Bringing embodied carbon upfront. Coordinated action for the building and construction sector to tackle embodied carbon.* World Green Building Council, 35 p.

Yu, J., Mishra, D.K., Wu, C. and Leung, C.K.Y. (2018a). Very-high-volume fly ash green concrete for applications in India, *Waste Management & Research*, 36, 520–526.

Yu, J., Lu, C., Leung, C.K.Y., and Li, G. (2017). Mechanical properties of green structural concrete with ultra-high-volume fly ash, *Construction and Building Materials*, 147, 510–518.

Yu, J., Yao, J., Lin, X., Li, H., Lam, J.Y.K., Leung, C.K.Y., Ivan M.L.S. and Kaimin S. (2018b). Tensile performance of sustainable strain-hardening cementitious composites with hybrid PVA and recycled PET fibers. *Cement Concrete Research*, 107, 110–123.

Zunino, F., and Scrivener, K.L. (2020). Influence of Kaolinite Content, Limestone Particle Size and Mixture Design on Early-Age Properties of Limestone Calcined Clay Cements (LC3). In: Bishnoi S. (eds) *Calcined Clays for Sustainable Concrete. RILEM Book Series*, 25. Springer, Singapore, pp. 331–337.

3

Use of Industrial Waste Slag in the Development of Self-Compacting Concrete for Sustainable Infrastructures

K.P. Sethy
Government College of Engineering Kalahandi, Bhawanipatna, India

K. K. Sahoo
Kalinga Institute of Industrial Technology, Bhubaneswar, India

Biswajit Jena
DRIEMS, Cuttack, India

CONTENTS

3.1 Introduction 23
3.2 Industrial Slag 24
3.3 Self-Compacting Concrete 24
3.4 Experimental Program 24
    3.4.1 Materials 24
    3.4.2 Mix Design 24
3.5 Results and Discussions 25
    3.5.1 Fresh Properties 25
        3.5.1.1 Self-Compactability Properties 25
        3.5.1.2 T_{50} Flow Time and V-Funnel Time 25
    3.5.2 Blocking Ratio (L-box test) 26
    3.5.3 HRWR Demand 27
    3.5.4 Hardened Properties 27
        3.5.4.1 Compressive Strength 27
3.6 Conclusions 28
References 28

3.1 Introduction

Industrial slag is a by-product of the iron industry, has been used in Portland cement concrete for long time, and is widely accepted in the industry as a replacement for cement in amounts up to approximately 20% by mass. Incorporating industrial slag reduces the environmental hazardous effects of concrete and, at the same time, consumes an otherwise unused waste product and by substituting for cement, which is the most polluting and energy exhaustive component of concrete (an estimated 5% of all manmade CO_2 emissions are due to cement production). Industrial slag also enhances the properties of concrete in numerous ways: greater strength, improved fluidity due to more paste volume, improved dimensional stability, more dense concrete, and increased durability of concrete structures.

It is unavoidable upon leaders of research in infrastructure, construction, and materials science to highlight the benefits of industrial slag, to facilitate its maximum use, and to transfer its importance to the

DOI: 10.1201/9781003217619-4

broader community. Hence, the objective of this proposed research is to reduce the environmental impact of concrete construction through extensive use of innovative materials and production techniques, resulting in dramatic increases in the percentage of industrial slag used in general construction.

3.2 Industrial Slag

Industrial slag is used to enhance many concrete properties though its particles are smoother, spherical, and smaller as compared to cement particles (industrial slag has a surface area about 430 m^2/kg compared to 370 m^2/kg for ordinary Portland cement). This leads to make the concrete denser, more flowable without any segregation and bleeding. Industrial slag has pozzolanic properties—it reacts with calcium hydroxide, an undesirable by-product of the cement hydration process—to produce calcium silicate hydrate and is the most needed cementitious product of cement hydration. This process increases strength at later ages, condenses heat of hydration, and decreases the rate of concrete shrinkage, resulting in a fall in thermal shrinkage cracking. Industrial slag has also been found to upgrade the pore structure of concrete and drop its permeability due to its fine structure and has good inferences on durability and strength. Even though its countless compensations, however, cement replacement with industrial slag is usually limited to 20–30% by mass in most commercial practice.

3.3 Self-Compacting Concrete

Compaction techniques are still complicated in areas of the structure with congested reinforcement and complex forms (ACI, 1987) [1]. SCC is a flowing concrete that can easily pass through reinforcement without the aid of external energy input. Its production and application symbolise one of the most significant developments in the area of concrete composites [2, 3, 11]. SCC is developed using the same basic ingredients that are used for producing the regular conventional concrete but by carefully modifying the mix proportions. The fillers which are most widely used to increase the viscosity are fly ash, slag, glass filler, limestone powder, silica fume and quartzite filler. Due to high workability requirement, a high-range water-reducing admixture (HRWRA) is mandatory in the development of SCC mixtures to ensure that concrete is able to flow under its own weight [4].

3.4 Experimental Program

3.4.1 Materials

An ordinary Portland cement of strength class 53 MPa complying with IS: 12269 (53 grade) [5] and slag meeting the requirements of ASTM C 618 were used as cementitious materials. The properties of cement and slag used in the present investigation were presented in Table 3.1. Four types of locally available aggregates: i.e., 20 mm aggregate, 12 mm aggregate, 6 mm aggregate, and fine aggregate of maximum size 4.75 mm, were mixed together in different proportions. Commercially available polycarboxylate-based superplasticiser complying with ASTM C 494 Type F requirements was used.

3.4.2 Mix Design

The mix proportions of SCC with different percentage replacements of slag are tabulated in Table 3.2. A total of five slag SCC mixtures were designed. The mixtures were designed as per the methodology described earlier for different strength grades and replacements of slag by considering the efficiency of slag (k) (Dinakar et al., 2013) [6]. To have a complete understanding, the widest possible range of concrete strengths was designed from 20–100 MPa with slag replacements varying between 30–90%. According to the mix design the high-strength mixes contain a lower percentage of slag whereas the low-strength mixes contain high-volume replacements of slag as described in Table 3.2.

TABLE 3.1

Composition of Cement and Slag

(%)	Cement	Slag
SiO_2	32.9	33.1
Al_2O_3	5.7	16.6
Fe_2O_3	3.9	0.6
CaO	62.5	34.8
MgO	1.2	8.0
Na_2O	0.1	0.2
K_2O	0.39	0.5
SO_3	2.4	0.4
LOI (loss of ignition)	1.2	0.3
Blaine surface area (m^2/kg)	370	430
Specific gravity	3.15	2.93

TABLE 3.2

Mix Proportions of the SCCs Investigated

Conc. Grade [MPa]	Conc. Name	Total Powder [kg/m³]	Slag [%]	k_{28}	Aggregate, [kg/m³]				[w/(c + k_{28} * g)]	HRWR [%]	VMA [%]
					20 mm	12 mm	6 mm	Sand			
20	SCC20	550	90	0.69	268	373	224	766	0.78	0.3	0.25
30	SCC30	550	80	0.70	271	376	227	774	0.59	0.5	0.30
60	SCC60	550	60	0.77	274	380	229	782	0.36	1.0	0.20
90	SCC90	550	40	0.92	282	392	236	806	0.27	1.5	0.20
100	SCC100	550	30	1.02	288	400	241	823	0.25	2.2	0.05

3.5 Results and Discussions

3.5.1 Fresh Properties

3.5.1.1 Self-Compactability Properties

Replacements of slag in high volumes had shown a remarkable influence on the fresh SCC properties. Figure 3.1 shows the slump flow test where one can see the well distribution of the aggregates throughout the concrete. From Table 3.3, it was noticed that as the slag replacement increases the corresponding slump flow also increases. The slump flow of SCC100 at 30% replacement of slag was observed to be 640 mm while this could be increased to 795 mm when the slag replacement increased to 90%. This may be due to the fact that the agglomeration of cement particles can be dispersed by slag particles (Nehdi et al., 2004) [7]. When cement is being replaced by slag, it is evident that a low amount of superplasticiser is needed and an increase in the quantity of water is required to maintain the desired self-compacting property. The broken cylindrical test specimens were used to determine the uniformity of the distribution of coarse aggregate visually which indicate proper segregation resistance of mixtures.

3.5.1.2 T_{50} Flow Time and V-Funnel Time

The measured T_{50} flow times were in the range of 4.66–8.66s. The result from Table 3.3 shows, as the incorporation of slag in high volumes in SCC mixes increases, the flow time decreases. 90 % and 30% slag replacements exhibited the lowest and the highest flow times.

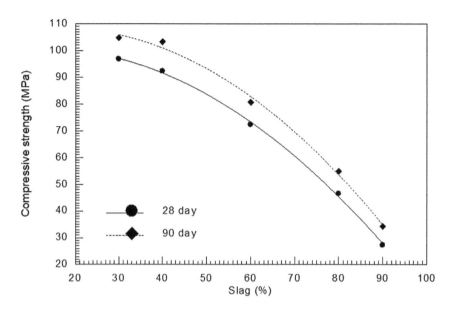

FIGURE 3.1 Slump flow test and measurement.

TABLE 3.3
Self-Compacting Properties of the Tested Mixes

Concrete Grade [MPa]	Concrete Name	T_{50} [sec]	Slump Flow [mm]	V-Funnel Flow Time [sec]	L-Box Ratio
20	SCC20	4.66	795	18.9	0.95
30	SCC30	5.21	785	19.21	0.89
60	SCC60	6.32	755	20.0	0.77
90	SCC90	8.57	715	24.31	0.74
100	SCC100	8.66	640	28.0	0.65

The results indicated that, the V-funnel time decreases with increasing slag replacement ratio, and they were in the range of 18.9–28 s, respectively. According to EFNARC guidelines (EFNARC, 2005), the upper limit for V-funnel flow is 25s. Even though the V-funnel flow time of SCC100 surpassed the upper limit; however, it still filled the concrete moulds by its own weight. It has been seen that, irrespective of effective water cementitious materials ratio, the V-funnel times decrease with increasing slag content. For example, SCC20 having a slag replacement of 90% has a V-funnel time of 18.9 s and this has been increased to 28s for SCC100 for slag replacement of 30%.

3.5.2 Blocking Ratio (L-box test)

The L-box ratio test indicates the passing ability of the SCC and is sensitive to blocking. The blocking ratio of SCC mixes varied from 0.65 to 0.95 for slag replacements varying between 30 to 90 % and illustrated in Table 3.3. From the results of the blocking ratio test, it can be observed that all the slag SCC mixtures were exhibiting a satisfactory blocking ratio. Even though for some of the mixes the ratios were found to be lower than that of the EFNARC guidelines [8], there was no tendency of blockage between the rebars. This supports the view reported earlier by Felekoglu et al. (2007) [9] that even though the blocking ratio is greater than 0.6, SCC still has been acknowledged to achieve good filling ability.

3.5.3 HRWR Demand

From the results, it is evident that as the slag content increases the corresponding demand for HRWR decreases for achieving the desired self-compacting properties. This could be due to the fact that the spherical nature of slag particles acts as a lubricant, the HRWR used does not react with the slag particles producing a repulsive force and finally the action of the superplasticisers will be only with the cement particles (Wattanalamlerd & Ouchi, 2005) [10].

3.5.4 Hardened Properties

3.5.4.1 Compressive Strength

Compressive strength test was performed at different ages, such as 3, 7, 28 and 90 days, and the results are presented in Table 3.4. Table 3.4 shows the variation of compressive strength with respect to the age for all percentage replacements. Furthermore, the inclusion of slag in SCC results in a significant improvement in the strength for various replacements studied. This may be attributed to the pozzolanic reaction of slag with the calcium hydroxide liberated during cement hydration.

It is obvious from the above investigation that by tailoring the replacement percentages of slag the maximum strength of SCC can only be obtained only at a particular level of replacement. Moreover, it was observed that low-strength SCCs can be developed only at high-volume slag replacements and high-strength SCCs are only possible at low percentage of replacement. Using the present experimental data, the strengths obtained at 28 and 90 days were plotted with respect to the slag replacement as shown in Figure 3.2. It can be seen that the variation of strengths at different probable replacements of slag for a specific strength was very minimal. This portrays the restrictions on the maximum replacement of slag

TABLE 3.4

Strength Properties of the Tested Mixes

Strength Grade [MPa]	Name	Compressive Strength [MPa]			
		3 days	7 days	28 days	90 days
20	SCC20	16.53	19.07	27.27	34.22
30	SCC30	25.52	37.86	46.5	54.98
60	SCC60	25.33	54.17	72.33	80.79
90	SCC90	33.93	64.84	92.28	103.3
100	SCC100	47.59	66.36	96.83	104.8

FIGURE 3.2 Compressive strength with respect to slag content.

for any specific strength. Lastly, from the present experimental investigation the slag replacement for developing the desired strength of SCC can easily be predicted.

3.6 Conclusions

The incorporation of industrial slag in high volumes was found to be highly beneficial in improving the self-compactibility and rheological properties of self-compacting concrete. Based on the results derived in this investigation the following conclusions can be drawn:

- All the high-volume slag SCCs has satisfied the EFNARC regulations in terms of the fresh properties. The developed SCCs were having good flowability, passing ability and highly resistant to segregation.
- The experimental investigations on SCC with slag using the earlier established cementing efficiency factors show that lower strengths such as 20 & 30 MPa can easily be produced at a slag replacement of about 80–90%, while high strengths of about 60–100 MPa can be developed with 30–60% slag replacement.
- The addition of slag in high volumes decreased the demand of HRWR in SCC mixes. As the slag content increased from 30 to 90% the HRWR demand decreased significantly from 2.2% to 0.3%.

REFERENCES

1. ACI Committee, 309 1987. Guide for Consolidation of Concrete, *ACI Materials Journal*, 84, 5, September–October, pp. 410–449.
2. Sideris, K.K., Manita, P., 2013. Residual mechanical characteristics and spalling resistance of fiber reinforced self-compacting concretes exposed to elevated temperatures. *Construction and Building Materials*, 41, 296–302.
3. Sua-Iam, G., Makul, N., 2013. Utilization of limestone powder to improve the properties of self-compacting concrete incorporating high volumes of untreated rice husk ash as fine aggregate. *Construction and Building Materials*, 38, 455–464.
4. Okamura, H., Ouchi, M., 2003. Self-compacting concrete. *Journal of Advanced Concrete Technology*, 1, 5–15.
5. IS: 12269. 1987. Specification for 53 grade ordinary Portland cement: BIS, New Delhi, 110002.
6. Dinakar, P., Sethy, K.P., Sahoo, U.C., 2013. Design of self-compacting Concrete with ground granulated blast furnace slag. *Journal of Materials & Design*, 43 (1), 161–169.
7. Nehdi ML., Pardhan M., Koshowski S., 2004. Durability of self-consolidating concrete incorporating high-volume replacement composite cements. *Cement and Concrete Research*, 34 (11):2103–2112.
8. European Guidelines for Self-Compacting Concrete. BIBM, CEMBUREAU, EFCA, EFNARC and ERMCO, Available online from: http://www.efnarc.org.2005.
9. Felekoglu B., Turkel, S., and Baradan, B. 2007. Effect of water/cement ratio on the fresh and hardened properties of self-compacting concrete. *Building Environment* 42:1795–1802.
10. Wattanalamlerd, C., and Ouchi M., 2005. *Flowability of fresh mortar in self-compacting concrete using fly ash*. In: Yu, Z., Shi, C., Khayat, K.H., Xie, Y., eds. Paris: RILEM Publication SARL; p. 261–270.
11. Jena, B., Mohanty, B.B., and Sahoo, K., 2019. Comparative study on self-compacting concrete reinforced with different chopped fibers, *Proceedings of the Institution of Civil Engineers - Construction Materials 2018* 171:2, 72–84

4 Influence of Functionally Graded Region in Ground Granulated Blast Furnace Slag (GGBS) Layered Composite Concrete

Sangram K. Sahoo and Benu Gopal Mohapatra
KIIT Deemed to be University, India

Sanjaya Kumar Patro
VSS University of Technology, India

Prasanna K. Acharya
KIIT Deemed to be University, India

CONTENTS

4.1 Introduction	29
4.2 Literature Review	30
4.3 Aim of the Study	30
4.4 Experimental Program	30
4.4.1 Materials	30
4.4.2 Mix Proportion	30
4.4.3 Specimens Preparation	31
4.5 Procedure	31
4.6 Results and Discussion	32
4.6.1 Compressive Strength	32
4.6.2 SEM and EDX Analysis	32
4.7 Conclusion	35
References	35

4.1 Introduction

Engineers and scientists are trying hard to improve the strength of concrete with new concepts to reduce cost without compromise in properties. Production of functionally graded concrete (FGC) with GGBS is one possible process to achieve more strength and durability in concrete. The functionally graded material (FGM) boasts two benefits, one is economy and the other is strength gain with the utilization of waste materials. The functionally graded region is formed due to co-extrusions on the involvement of multiple cement-based layers at the face of their interface. The concrete layers of FGC having different properties with more interfaces attained an intricate lamination property. The act of reducing stresses within FGC specimens happens due to the force of resistance offered by the molecules and so formed lamination property.

DOI: 10.1201/9781003217619-5

4.2 Literature Review

Every specimen of different volume fraction of normal concrete and fly ash concrete gives an asymmetric non-linear behavioral result in the elastic modulus (Bajaj et al., 2014). When the composition changes in the layer, the stress concentration is the main advantage attained by the structural element over conventional composites (Nazari & Sanjayan, 2014). FGM components in the formation of FGC specimens give beneficial results due to the change of graded regions with the provision of appropriate compositions with appropriate materials, which will be homogeneous with laminate composite and will not delaminate during crack propagation (Zhang et al., 2017). The FGC layer lining possesses more elasticity value than a normal graded layer lining (Chen et al., 2008). The elastic property of materials used in FGM gives a higher value in comparison with normal concrete (Nazari & Sanjayan, 2015). In FGC, a hybrid fiber reinforcement layer at the top offered the best impact resistance, high ultra-high performance steel fiber reinforcement in the middle showed excellent performance against penetrations and steel fiber at the bottom layer had the best performance against spalling resistance (Lai et al., 2019). FGC saves 40% binding material, achieves the desired strength, controls the post-fracture effect, reduces the progression of corrosion, provides insulation against heat, and reduces energy consumption and carbon emission (Torelli et al., 2020).

4.3 Aim of the Study

The present study aims to enhance the compressive strength and evaluation of mechanical properties of composite concrete having two layers of cement concrete with different binding constituents.

4.4 Experimental Program

4.4.1 Materials

The GGBS used in this study was collected from an operating plant of Neelachal Ispat Nigam, Duburi, Odisha, India. The coarse aggregate (CA) conforming to IS 383 (1970) was collected from local crushers. The local river sand that confirmed the requirements of zone-III of IS 383 (1970) was used as fine aggregate (FA). OPC-43 grade that confirmed the requirements IS 8112 (1989) was used as binding material. The chemical composition of OPC and GGBS is presented in Table 4.1.

4.4.2 Mix Proportion

The concrete mixes of 0, 20, 35 and 50% GGBS were prepared. These mixtures were named M_0, M_1, M_2 and M_3. The details of the ingredients of these mixes are presented in Table 4.2.

TABLE 4.1

Chemical Composition of OPC 43 and GGBS

Material	SiO_2	Al_2O_3	Fe_2O_3	C_aO	SO_3	Na_2O	K_2O	M_gO	Other Oxides
					%				
GGBS	32.384	19.489	1.406	33.682	0.679	0.248	0.908	9.835	0.924
OPC (43)	21.07	5.65	4.05	63.24	2.15	0.20	0.45	1.16	---
As per IS-8112 (1989) for OPC-43	17–25	3–8	0.5–6	60–67	2–3.5	0.3–1.2		0.5–4	---

TABLE 4.2

Ingredients of Concrete Mixes

Mix Designation	M_0	M_1	M_2	M_3
OPC-43 (kg/m³)	340	272	221	170
GGBS (kg/m³)	--	68	119	170
Fine aggregate (kg/m³)	718.48	718.48	718.48	718.48
Coarse aggregate (kg/m³)	1,203.93	1,203.93	1,203.93	1,203.93
Water/cement ratio	0.5	0.5	0.5	0.5
GGBS (%)	0	20	35	50

4.4.3 Specimens Preparation

The concrete specimens were prepared based on the mix proportion of Table 4.2 and used as reference concrete with designation M_0, M_1, M_2 and M_3. Then composite concrete specimens were prepared with two layers of concrete having 75mm thickness each. One layer was prepared from OPC and another layer from different percentage combinations (20, 35 and 50%) of GGBS. Those specimens were designated as S_1, S_2 and S_3 respectively. The FGC specimens S_1 was made of 75mm thick OPC and 75mm of GGBS blended concrete with 20% GGBS content. Similarly, S_2 contained 75mm of OPC and 75mm of GGBS blended concrete with 35% GGBS content and S_3 contained 75mm of OPC and 75mm of GGBS blended concrete with 50% GGBS content.

4.5 Procedure

Based on the above mix design the compressive strength test of 150mm cubes was carried out as per IS: 516–1959. The test was conducted as shown in Figure 4.1, configuration 1 (C_1) and configuration 2 (C_2). The load was applied as perpendicular and parallel to the interface of layers. Figure 4.1 shows the details of configurations of the specimens considering as two-element specimens cast by pouring of M_0 and M_1/M_2/M_3 mixtures in two layers of 75mm thickness. In the S_1C_1 configuration, the M_0 mixture is of 75mm and M_1 is 75mm. The FGC specimen S_1 in C_1 configurations is named as C_1S_1. Similarly, S_2 and S_3 in C_1 configuration are termed as C_1S_2 and C_1S_3. Further, the S_1 in C_2 is named as S_1C_2. In the same way, S_2 and S_3 in C_2 configuration are named as C_1S_2 and C_1S_3.

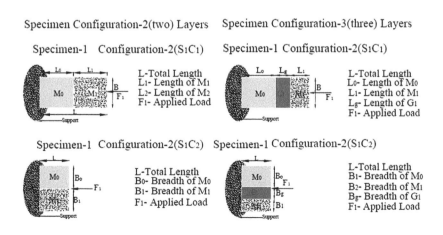

FIGURE 4.1 Details in configurations 1 and 2 of specimen 1.

4.6 Results and Discussion

4.6.1 Compressive Strength

The crushing strength of one-layered concrete made of OPC (M_0), GGBS blended concrete (M_1, M_2 and M_3) tested at 7, 14, 21 and 28 days are presented in Figure 4.2. The same of double-layered concrete, the first layer being made of OPC and the other of GGBC blended concrete are presented in Figure 4.3. The crushing strength value of FGC specimens gave different results in different configurations. Irrespective of configurations (C_1 and C_2), the FGC specimens gave more crushing strength value than the M_0 specimen. The crushing strength values at the different configurations of specimens S_1 to S_3 gave different values in C_1 and C_2 configurations. All crushing strength values of the C_2 configuration for all FGC specimens were found higher than the C_1 configuration of S_1 to S_3 FGC specimens and normal specimens of M_0 to M_3. It is observed that in FGC specimens a middle layer between two layers has been generated and herein named as a graded layer. Due to the presence of this graded layer, the crushing strength value was observed more in C_1 and C_2 configurations.

4.6.2 SEM and EDX Analysis

Figures 4.4, 4.5 and 4.6 presented the microstructures of M_0, M_2 and graded region of the S_2 monolithic FGC specimen. As per visual observation, it was seen that there exists a very clear boundary between the two layers. One layer was off-white and the other was bluish-white. One sample for the microstructural study was collected from the boundary of these two layers for a width of 40mm × 40mm size. The other two samples were collected apart from this middle layer, which is from the bluish-white surface, and off white surface, but from the boundary of the middle sample area. All samples are cleaned by rubbing with

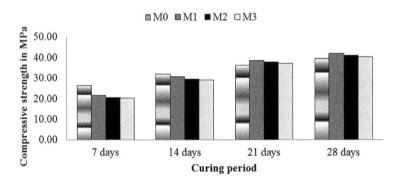

FIGURE 4.2 Compressive strength of M_0 to M_3 specimens.

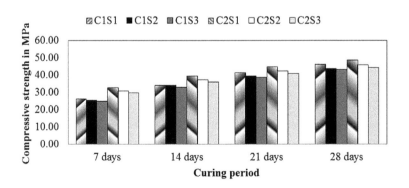

FIGURE 4.3 Compressive strength of C_1S_1 to C_2S_3 specimens.

Influence of Functionally Graded Region 33

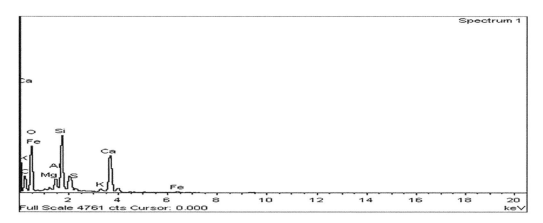

FIGURE 4.4 EDS analysis of M₀.

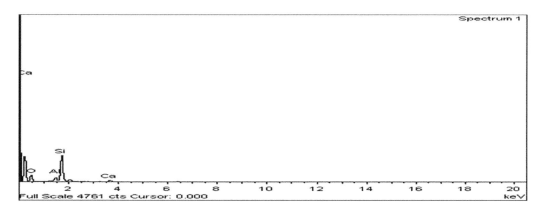

FIGURE 4.5 EDS analysis of Graded layer.

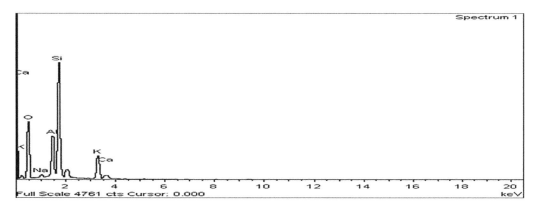

FIGURE 4.6 EDS analysis of M₂.

hard stone and dried in 100°C for 24hrs, and then are coated with platinum for examination under SEM microscope. The Si and Al peak intensity was calculated from EDS analysis from 9 number of points from 3 samples collected. The 50-micron imaginary parallel lines had been drawn in several intervals for conducting EDS analysis on the above specimens and along each line, 5 points have been selected and analysis carried out with 1000μm telescopic magnification of GGBS-blended concrete layer and normal concrete layer. The graded layer was analyzed with 3000μm telescopic magnification. The functionally

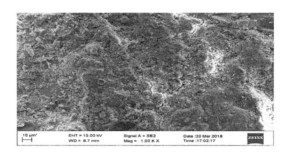

FIGURE 4.7 SEM analysis of OPC concrete layer (M_0).

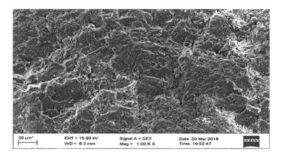

FIGURE 4.8 SEM analysis of concrete layer with GGBS (M_2).

FIGURE 4.9 SEM analysis of graded layer.

graded region boundary was considered as per the mean of Si and Al peak difference with 5% as compared with the other two layers collected from the monolithic concrete specimen.

Figures 4.7, 4.8 and 4.9 represent the microstructure of M_0 and M_2 and graded layers. The microstructure of M_0 and M_2 shows fracture surfaces with the brittle phenomenon. This phenomenon is not present in the graded layer. The presence of gradual changes in brittle structure surface in the layer M_0 and M_2 towards graded layer reflected in this monolithic specimen. There are no brittle fractures in the graded region and the surface is showing as a plane structure. Hence the gradual change in fractures towards the graded layer is indicating the changes in their properties gradually. The depth of the graded layer is counted to be 19.5mm through EDS analysis as shown in Figure 4.10.

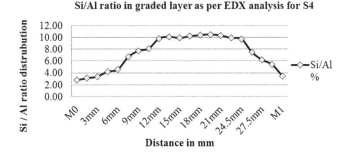

FIGURE 4.10 EDS analysis of Si/Al ratio of graded layer.

4.7 Conclusion

It is concluded from the above study that the FGC specimens offered more compressive strength in comparison to normal concrete specimens due to the generation of a graded layer at the place of the interface of layers. The compressive strength is found more when the thickness of the graded layer is more. The FGC specimen containing a 75mm thick layer of blended concrete made of 20% GGBS and another layer of same thickness concrete made with OPC offered the highest compressive strength. The existence and thickness of the graded layer are validated and calculated through SEM and EDX analyses.

REFERENCES

Bajaj, K., Shrivastava, Y., Dhoke, P. 2014. Experimental study of functionally graded beam with fly ash. *Journal of Institution of Engineers (India) Series-A*, 94(4), 219–227.

Chen, Y., Struble, I. J., Paulino, G.H. 2008. Fabrication of Functionally graded – cellular structures of cement-based materials by co-extrusion. AIP Conference Proceeding, 973, 532.

IS, 383. 1970, Reaffirmed 2002. Specifications for coarse and fine aggregates from natural sources for concrete. Bureau of Indian Standards, New Delhi, India.

IS, 8112. 1989. 43 Grade ordinary portland cement –specifications. Bureau of Indian Standards, Delhi, India.

Lai, J., Yang, H., Wang, H., Zheng, X., Wang, Q. 2019. Penetration experiments and simulation of three-layer functionally graded cementitious composite subjected to multiple projectile impacts. *Construction and Building Materials* 196, 499–511.

Nazari, A., Sanjayan, J.G. 2014. Modeling of fracture strength of functionally graded geopolymer. *Construction and Building Materials*, 58, 38–45.

Nazari, A., Sanjayan, J.G. 2015. Compressive strength of functionally graded geopolymers: Role of position of layers. *Construction and Building Materials*, 75, 31–34.

Torelli, G., Fernandez, M.G., Lees, J.M. 2020. Functionally graded concrete: Design objectives, production techniques and analysis methods for layered and continuously graded elements. *Construction and Building Materials*, 242, 118040.

Zhang, N., Lu, A., Li, C.C., Zhou, J., Zhang, X., Wang, S., Chen, X. 2017. Support performance of functionally graded concrete lining. *Construction and Building Materials*, 147, 35–47.

5

Utilization of Fly Ash as a Replacement of Sand in Concrete for Sustainable Construction

Ankita Sikder
KIIT Deemed to be University, India

Sukamal Kanta Ghosh
Greater Kolkata College of Engineering and Management, India

CONTENTS

5.1 Introduction 37
5.2 Literature Review 38
5.3 Materials and Methods 38
    5.3.1 Cement 38
    5.3.2 Aggregates 39
        5.3.2.1 Fine Aggregate 39
        5.3.2.2 Coarse Aggregate 39
        5.3.2.3 Fly Ash 39
    5.3.3 Mix Proportions 40
    5.3.4 Preparation and Casting of Test Specimens 40
5.4 Test Results and Discussion 40
    5.4.1 Fresh Properties 40
    5.4.2 Compressive Strength 41
    5.4.3 Split Tensile Strength 41
5.5 Conclusions 42
References 42

5.1 Introduction

The management of solid waste is one of the biggest headaches for developing countries like India. Fly ash is one of the major solid wastes produced during burning of coal, which leaves ashes behind it. Some ash falls to the bottom-most part of the furnace (called bottom ash) and some ash is lifted by the hot burning gases of the furnace (called fly ash) and separated by collection devices (Aggarwal & Siddique, 2014). The thermal power plants in India are generating nearly 213.188 million tonnes of fly ash each year (2016) and is projected to exceed 907.1 million tonnes by the year 2032. Fly ash is already being used as cement replacement in construction industry, yet 40–50% of the fly ash remains unutilized (Yadav & Fulekar, 2018). On the other hand, excessive utilization of river sand as fine aggregate in concrete based construction is ruining Mother Nature. These two different problems can be minimized through a single approach by utilizing fly ash in concrete as a fine aggregate.

DOI: 10.1201/9781003217619-6

5.2 Literature Review

Several numbers of investigations have already been done to utilize fly ash in concrete as cement substitute, but there exist only a few literatures on utilization of fly ash as partial or full replacement of sand.

Aggarwal et al. researched about the outcome of waste foundry sand and bottom ash in equal amount of sand was partially replaced in different percentages (0–60%), on durability characteristics and mechanical characteristics of concrete accompanied by micro-structural analysis. Test results revealed that up to 50% substitution of sand, strength of concrete increases more than conventional concrete because after that it leads to flaws in concrete. Optimum replacement percentage was 30% which forms large C-S-H gel resulting in development of denser micro-structure.

Christy & Tensing (2010) investigated concrete's compressive strength where fly ash was utilized as both partial substitution for cement and sand. Results from those tests showed substantial enhancement in strength because of pozzolanic activity of fly ash leads to the densification of the mix.

Güneyisi et al. (2015) explored the consequence of lightweight fly ash fine aggregates (LWFA) as part-substitution in self-compacting mortars in different percentages (0–100%). Tests were done to find out fresh and harden properties of concrete and results showed that with increase in percentage of LWFA the flowability and the workability of mortars also increases but the permeability and strength decreases. The spherical shape of the LWFA was the main reason behind the good flowability and workability. On the other hand, because of higher water absorption rate and lower specific gravity of LWFA there would be a decline in compressive strength. Although, decreasing rate of compressive strength decreases over time due to pozzolanic activity of LWFA.

Siddique (2003) executed experiments to assess the workability, modulus of elasticity, flexural strength, compressive strength and split tensile strength of concrete where fly ash has been utilized as a partly substitute of sand (10%, 20%, 30%, 40%, and 50% by weight) and found that workability decreased, modulus of elasticity and strength increased. This strength gain resulted from improved interfacial bond between aggregates and pastes and pozzolanic activity of fly-ash.

Kockal & Ozturan (2011) used design expert software to investigate specific gravity, crushing strength and water absorption values of concrete. Where, fly ash was utilized with and without the combination of glass powder and bentonite as partial replacement to lightweight aggregate and achieved high strength lightweight concrete. It was seen that specific gravity and crushing strength tends to increase with the use of low binder content at high temperature. Water absorption decreased as the temperature increases as a result of relatively impervious pores obtained after the melted liquid phase. Nevertheless, this type of binder has unnoticeable effect on water absorption and specific gravity.

5.3 Materials and Methods

5.3.1 Cement

The cement used was OPC-43 grade and tested as per IS: 8112-1989. Table 5.1 shows the characteristics of cement.

TABLE 5.1

Properties of Cement

Properties		Values
Consistency		30%
Setting time	Initial	49 mins
	Final	486 mins
Compressive strength	3 Day	24 MPa
	7 Day	35 MPa
	28 Day	47 MPa

Utilization of Fly Ash 39

5.3.2 Aggregates

5.3.2.1 Fine Aggregate

Natural sand (max. size 4.75mm) was taken as fine aggregate and tests are done as per IS: 383-1970.

5.3.2.2 Coarse Aggregate

Gravel with 20mm nominal size used as coarse aggregates and tests are done as per IS: 383-1970. All characteristics of aggregates are shown in Table 5.2 and 5.3.

5.3.2.3 Fly Ash

Fly ash (Class F) was acquired from Kolaghat Thermal Power Plant (KTPP) in West Bengal. As per ASTM C 311, the chemical composition was determined and shown in Table 5.4.

TABLE 5.2

Physical Properties of Aggregate

Properties	Sand	Coarse Aggregate
Sp. Gravity	2.64	2.6
Fineness modulus	7.2	2.8
Unit (kg/m^3)	1639	1695

TABLE 5.3

Sieve Analysis of Aggregate

Fine Aggregate		Coarse Aggregate	
Sieve Size	Percentage Passing	Sieve Size	Percentage Passing
4.75 mm	99.3	40 mm	98.8
2.36 mm	92.1	20 mm	32
1.18 mm	70.8	4.75 mm	1
600 μm	60.5		
300 μm	20.4		
150 μm	6.8		

TABLE 5.4

Chemical Composition of Fly Ash

Oxides	Percentage by Mass
SiO_2	49.5
Al_2O_3	28.3
Fe_2O_3	4.3
CaO	5.3
MgO	1.8
TiO_2	1.1
Na_2O	0.4
K_2O	0.9
SO_3	1.8
LOI (loss on ignition)	2.1
Moisture	0.4

TABLE 5.5

Details of the Mix Proportion (kg/m³)

Mix	Cement	Water	Sand	Fly Ash	Aggregates
MF0	450	180	450	Nil	900
MF10	450	180	405	45	900
MF20	450	180	360	90	900
MF25	450	180	337.5	112.5	900
MF30	450	180	315	135	900
MF35	450	180	292.5	157.5	900
MF40	450	180	270	180	900

5.3.3 Mix Proportions

Seven mixture proportions were made. Out of these, six mixes are prepared with different fly ash content (10%, 20%, 25%, 30%, 35%, and 40%) and the remaining one is plain concrete with 1:1:2 (cement: fine aggregate: coarse aggregate) ratio. Fly ash (by weight) was used as a replacement of sand. The standard plain concrete mix which has no fly ash content was equipoised as per IS: 10262-2009 to acquire 28 days compressive strength of 49.7 MPa in cube samples. Mixes of concrete were prepared in a power-driven rotating type drum mixer having capacity of 0.76m³. Detailed specifications of these mixture proportions are stated in the Table 5.5.

5.3.4 Preparation and Casting of Test Specimens

To perform test for mechanical strengths, concrete cubes (150mm) and cylinders (150 × 300mm) were cast to find the compressive strength and splitting tensile strength respectively. After casting, by using a steel trowel, all test samples were smoothed. After finishing, with a plastic sheet, the samples were wrapped immediately to minimize the loss of moisture. After that, the test specimens were kept in the molding room at a temperature of around 25°C for 24h. After that this samples were unmolded and placed in a water curing tank.

5.4 Test Results and Discussion

5.4.1 Fresh Properties

To find workability of mixes, slump cone test was done. Figure 5.1 shows the graphical representation of slump test. With the increasing amount of fly ash content workability decreases.

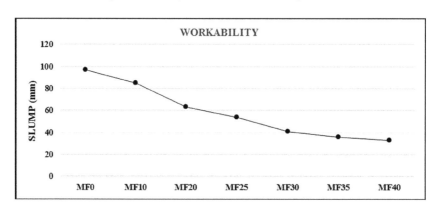

FIGURE 5.1 Slump of all mixes.

Utilization of Fly Ash

5.4.2 Compressive Strength

Compressive strength of concrete mixtures was figured out at 7, 28, 60 and 90 days of curing. These test outcomes are given in Figure 5.2. Higher values of compressive strength are seen for sand-substituted fly ash concrete mixes compared to the conventional concrete mix at different stages and with increasing amount of fly ash compressive strength is enhanced. The strength gain of concrete consisting fly ash demonstrates enhanced strength at earlier ages as a result of incorporation of fly ash as part-substitution of fine aggregate begins pozzolanic activity and makes the matrix of concrete denser, and as a result of this, even at initial times, the strength of fly ash concrete is higher compared to that of conventional concrete.

5.4.3 Split Tensile Strength

Concrete's split tensile strength was calculated at 7, 28, and 60 days of curing. Test results are shown in Figure 5.3. Split tensile strength of concrete when sand was replaced by fly ash showed greater values

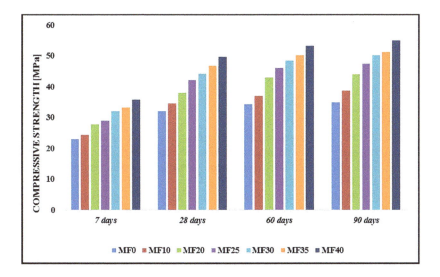

FIGURE 5.2 Compressive strength of all mixes.

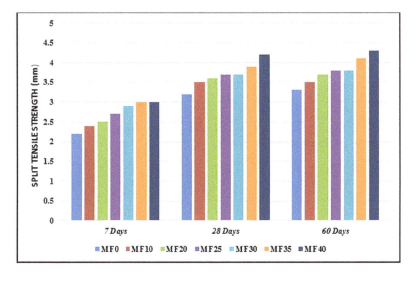

FIGURE 5.3 Split tensile strength of different mixes.

than ordinary concrete mix at different stages, and with the rise in fly ash quantity split tensile strength improves. The rate of improvement in strength was slower, initially (7 days) but after 28 days, it increased conspicuously. The late pozzolanic activity may be the cause of slower development of pozzolanic C–S–H gel.

5.5 Conclusions

After analyzing all the data that we got from test results, conclusions can be drawn:

1. Fly ash (Class F) could be utilized by partly substituting sand in structural concrete.
2. Compressive and split tensile strength of plain concrete is significantly lower when collated to the fly ash concrete replaced by fine aggregate.
3. By increasing fly ash content compressive and split tensile strength of fly ash concrete replaced by sand rises.
4. With age mechanical strength of concrete continued to increase where sand was substituted by fly ash.
5. Workability is affected adversely in concrete where sand was substituted by fly ash. With increasing amount of fly ash as a substitute of sand workability continues to decrease.

REFERENCES

Aggarwal and Siddique. 2014. Microstructure and properties of concrete using bottom ash and waste foundry sand as partial replacement of fine aggregates, *Construction and Building Materials*, 54 pp. 210–223.

Christy and Tensing. 2010. Effect of class-F fly ash as partial replacement with cement and fine aggregate in mortar, *IJEMS*, 17, 140–144.

Güneyisi, E., et al., 2015 Utilization of cold bonded fly ash lightweight fine aggregates as a partial substitution of natural fine aggregate in self-compacting mortars, *Construction and Building Materials*, 74 pp. 9–16.

Islam I and Bin. 2018. Partial replacement of fine aggregate with fly ash and it's compressive strength, *IJEDR*, 6(1), 383–385.

Kockal and Ozturan. 2011. Optimization of properties of fly ash aggregates for high-strength lightweight concrete production. *Materials and Design*, 32 3586–3593.

Pofale and Deo. 2010. A study of fine aggregate replacement with fly ash: an environmentally friendly and economical solution, 1–18.

Rajamane, N., et al. 2007. Prediction of compressive strength of concrete with fly ash as sand replacement material, *Cement & Concrete Composites*, 29, 218–223.

Saxena and Pofale. 2016. Effective utilization of fly ash and waste gravel in green concrete by replacing natural sand and crushed coarse aggregate, *Materials Today*, 4, 9777–9783.

Siddique, R.. 2003. Effect of fine aggregate replacement with Class F fly ash on the mechanical properties of concrete, *Cement and Concrete Research*, *33*, 539–547.

Thomas and Nair. 2015. Fly ash as a Fine Aggregate Replacement in Concrete Building Blocks. *IJEART,* 1(2), 47–51.

Yadav and Fulekar. 2018. The current scenario of thermal power plants and fly ash: production and utilization with a focus in India, *IJAERD*, 5(4) 768–777.

6

Properties of Concrete at Elevated Temperature Using Waste HDPE as Fibre and Copper Slag as Mineral Admixture

Marabathina Maheswara Rao and Sanjaya Kumar Patro
Veer Surendra Sai University of Technology, India

CONTENTS

6.1 Introduction 43
6.2 Literature Review 44
6.3 Methodology 44
6.4 Result and Discussions 45
    6.4.1 Slump Test 45
    6.4.2 Mechanical Properties 46
        6.4.2.1 Compressive Strength 46
        6.4.2.2 Split Tensile Strength 46
        6.4.2.3 Flexural Strength 47
6.5 Effect of Elevated Temperature 47
    6.5.1 Loss of Weight 47
    6.5.2 Colour Change and Appearance 47
    6.5.3 Compressive Strength 48
6.6 Effect of Elevated Temperature on Durability 48
    6.6.1 Sorptivity at Elevated Temperatures 48
    6.6.2 Water Absorption After Applying Elevated Temperature 49
6.7 Conclusions 50
References 50

6.1 Introduction

High density polyethylene (HDPE) is a plastic widely used for manufacturing of water pipes, chemical cans, ropes, and many more. HDPE is distinguished by its tensile property and outstanding durability, and it is utilised in roads and buildings. The strength attributes of plastic fibre reinforced concrete (FRC) are studied by many researchers (Rao, 2018; Pešić et al., 2016), but very few investigated on its attributes at high temperatures (Zheng, Li & Wang, 2012). As a fair confident material, it is desired to study the strength and durability properties of plastic FRC after applying elevated temperatures. It is essential to know the performance of concrete in compressive strength after elevated temperature in case of fire, in order to repair and evaluate construction succeeding fire (EN 1992-1-2, 2004). More pertinent investigations are done on the normal strength (Husem, 2006). The results revealed that mechanical properties diminished slowly with rising temperature, and explosive spalling appears

DOI: 10.1201/9781003217619-7

in high-strength concrete during the heating process (Kodur, 2004; Kalifa, Menneteau & Quenard, 2000). Spalling can be prevented and also the mechanical attributes boosted at elevated temperature by inclusion of the steel and polypropylene fibres in concrete (Pliya, Beaucour & Noumowé, 2011). High temperature can be effectively resisted by concrete containing steel and polypropylene fibres (Zheng, Li & Wang, 2012).

To investigate the compressive strength and microstructures of HDPE fibre reinforced concreted (HDPE-FRC) at elevated temperature exposure, elevated temperature tests, compression strength tests, as well as microstructure tests, were performed with the 50 × 50 × 50mm cubes by cutting 100 × 100 × 100mm cubes into four equal parts. The appearance and mass loss of HDPE fibre reinforced concrete were investigated in the heating process. The effect of fibre content and temperature on the compressive strength and durability were studied. Copper slag is being utilised for production of tiles, paving and tools for cutting and abrasing. If it is utilised in concrete, a huge quantity of copper slag waste can be removed, which also benefits construction cost reduction, low heat of hydration, and low permeability.

6.2 Literature Review

Brindha and Nagan investigated by replacing 0–20% cement with copper slag on M20 grade (Brindha & Nagan, 2011). Al-Jabri et al. studied replacing 5% cement with copper slag and other mixtures, with 13.5% copper slag, 1.5% cement by-pass dust and 85% Portland cement, with water-to-binder ratios of: 0.5, 0.6 and 0.7 (Al-Jabri et al., 2006). The flexural and compressive strengths were reduced at 30% cement replacement by copper slag just as to fly ash, the reduction is less after 28 days. The similar pozzolanic activity shown by copper slag and fly ash and more pozzolanic activity shown by silica fume (Isabel Sanchez et al., 2008). Antonio replaced 15% cement by mass of ground copper slag (Arino & Mobasher, 1999). Najimi and Pourkhorshidi replaced 5%, 10% and 15% of cement by copper slag (Najimi & Pourkhorshidi, 2011). Shi et al. reported that up to 10% of cement replacement with copper slag in mortar the leachability is within the standards. 5–7.5% by weight replacement with copper slag shown optimum strength, replacement of 10–15% cement clinker shown more abrasion resistance of the mortar (Brindha & Nagan, 2011; Shi, Meyer & Behnood, 2008).

6.3 Methodology

Till now, researchers investigated up to 15% cement replacement and 20–80% sand replacement in concrete. This work investigated replacing 60% of cement in HDPE fibre reinforced concrete.

Concrete of M30 grade was utilised to cast the specimens. Cubes of size 150mm × 150mm × 150mm, cylinders of size 150diameter × 300mm height, prisms of size 100mm × 100mm × 500mm were cast to test the compressive strength, split tensile strength, and flexural strength respectively. All the specimens were cured for 28 days, 56 days and 90 days and tested as per Indian code. Specimens were also cast with replacement of 60% of cement with copper slag as per the literature.

Cubes of size 100mm × 100mm × 100mm are casted and kept at 90°C in an accelerated curing tank for three days after 24 hours of casting, then the specimens are transferred to water curing tank for 56 days, then kept in air curing for two months (Zheng, Li & Wang, 2012). Specimens were made by cutting the 100mm × 100mm × 100mm cube in to four parts i.e., 50mm × 50mm × 100mm. and kept in muffle furnace of capacity 1200°C. Specimens are kept continuously in the muffle furnace for two hours after reaching target temperature at the rate of 4°C/min and cooled in the muffle furnace till reaches the room temperature. The HDPE fibre and copper slag are shown in Figure 6.1. The concrete mix with and without copper slag is shown in Table 6.1.

Properties of Concrete at Elevated Temperature 45

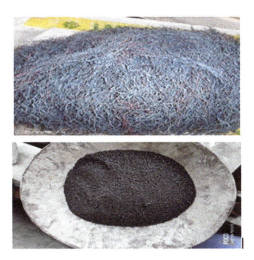

FIGURE 6.1 HDPE fibre and copper slag.

TABLE 6.1

Concrete Mix with and without Copper Slag

Mix Description	Mix with 0% Replacement of Cement (CS00)	Mix with 60% Replacement of Cement with Copper Slag (CS60)
0% HDPE	CS00H0	CS60H0
0.2% HDPE	CS00H2	CS60H2
0.4% HDPE	CS00H4	CS60H4
0.6% HDPE	CS00H6	CS60H6

6.4 Result and Discussions

6.4.1 Slump Test

The workability of CS00 and CS60 with various percentages of HDPE fibre is shown in Figure 6.2. The workability is increased at CS00H2 and then decreased continuously till CS00H6. The slump is increased by 18.18% and 15% at 0.2% HDPE comparing to 0% HDPE for CS00 and CS60 respectively. The maximum workability is obtained at 0.2% HDPE.

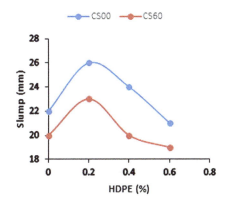

FIGURE 6.2 Slump (mm).

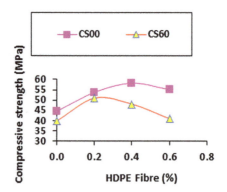

FIGURE 6.3 Compressive strength (MPa).

6.4.2 Mechanical Properties

6.4.2.1 Compressive Strength

Compressive strength with CS00 and CS60 with various percentages of HDPE fibre is shown in Figure 6.3. The graph shows that there is a continuous increase till 0.4% and 0.2% at 28 for CS00 and CS60 respectively. CS60 has shown low compressive strength comparing to CS00. Maximum compressive strength is observed at 0.2% HDPE with CS60 and at 0.4% with CS00. Comparing to CS00 and CS60 the compressive strength is decreasing at all percentages of HDPE fibre and there is increase in decrement with increase in fibre content.

It is concluded that the there is a 5.44% and 20.97% increasing in compressive strength at 0.2% and 0.4% with CS00 comparing CS60, respectively. It is concluded that the optimum compressive strength occurred at CS00H4 and at CS60H2, increased by 30.39% and 14.20% comparing to normal concrete.

6.4.2.2 Split Tensile Strength

Comparing concrete with and without copper slag the split tensile strength with CS00 and CS60 with various percentages of HDPE fibre at 90 days water curing is shown in Figure 6.4. From this figure the maximum split tensile strength is at 0.2% at 90 days for CS00 as well as CS60, but there is a reduction in split tensile strength comparing to CS60 in CS00. Comparing to CS00 the CS60 strength is decreasing at all percentages of HDPE fibre, and there is increase in decrement with increasing fibre content. It is concluded that there is 8.61% increase in split tensile strength at 0.2% with CS60 comparing to CS00. It is concluded that the optimum split tensile strength occurred at CS00H2 and at CS60H2, increased by 12.5% and 22.18% comparing to normal concrete.

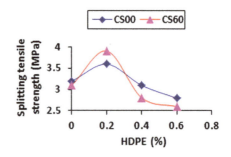

FIGURE 6.4 Split tensile strength (MPa).

Properties of Concrete at Elevated Temperature 47

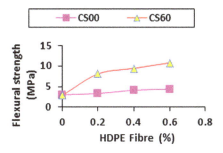

FIGURE 6.5 Flexural strength.

6.4.2.3 Flexural Strength

The flexural strength of 90 days water-cured specimens with and without copper slag is shown in Figure 6.5. The flexural strength is continuously increasing as increase in percentage of HDPE fibre in concrete. The graph shows CS60 gained more strength than CS00 at all percentages of HDPE fibre in concrete. It is concluded that the there is a 43.79% and 253.92% increasing in flexural strength at 0.6% HDPE with CS00 and CS60 respectively, comparing to normal concrete. It is concluded that the flexural strength is increased by 146.1% at CS60H6, comparing to CS00H6.

6.5 Effect of Elevated Temperature

6.5.1 Loss of Weight

The specimens were weighed before and after elevated temperature and a graph drawn with temperature versus loss of weight is shown as Figure 6.6. The weight of the specimens is measured with a 0.01g accurate weighing balance before and after elevated temperature. This test is also done with the specimens without copper slag. The loss of weight is increasing with increase in temperature. The maximum loss of weight occurred with CS00H4.

6.5.2 Colour Change and Appearance

The samples shown change in colour with high temperatures. Flaking and minor cracks appeared at 600°C and more cracks and flaking at 800–900°C.

The following may be the causes for no subjection spalling happened during the whole heating process:

1. The literature shows that the polypropylene fibres can enhance the compressive strength, tensile strength, and crack energy (Poon, Shui & Lam, 2004; Peng et al., 2006). Hence, HDPE fibres, perhaps, would be favourable to help concrete conquer a build-up of vapour pressure at elevated temperature and resulting non-development of spalling.

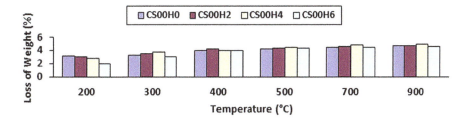

FIGURE 6.6 Loss of weight.

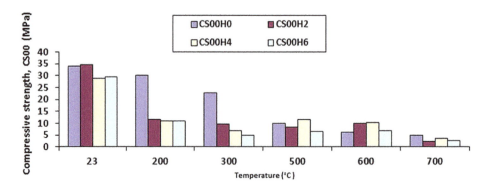

FIGURE 6.7 Effect of elevated temperature on compressive strength.

2. (2) HDPE fibres average melting point is 150 °C and generate channels to alleviate the internal vapour coercion of HDPE-FRC. This delivery of the vapour pressure appreciably lowers the spalling propensity of HDPE-FRC at elevated temperature, which is close with the thoughts of investigators (Peng et al., 2006; Kalifa, Chéné & Gallé, 2001; Tanyildizi, 2009). So, the spalling at elevated temperature of HDPE-FRC was avoided by the effect of the two reasons as said above.

The colour change to slate grey, greyish brown, darkish brown, greyish white and pinkish white at 200°C, 400°C, 500°C, 700°C, and 900°C, respectively.

6.5.3 Compressive Strength

After three days of application of elevated temperature the specimens were tested on a compression testing machine at the rate of loading 1–1.5kN/s and the compressive strength versus temperature graph is drawn as shown in Figure **6.7**. As the melting temperature of HDPE fibre is 120–180°C, the compressive strength is reduced even at 200°C due to melting of HDPE fibre. The high temperature resulted in the specimens colour change, spalling, cracks, meshy cracks and also lowered their strength.

6.6 Effect of Elevated Temperature on Durability

6.6.1 Sorptivity at Elevated Temperatures

Sorptivity at 23°C and 600°C are shown in Figure **6.8** and Figure 6.9 respectively. Sorptivity is increased with increase in HDPE fibre percentage whereas it is reduced for CS00H4. The lowest value of sorptivity

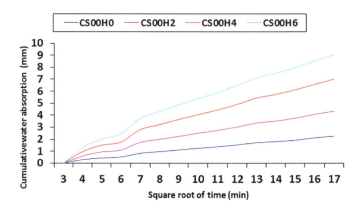

FIGURE 6.8 Sorptivity at 23°C.

Properties of Concrete at Elevated Temperature

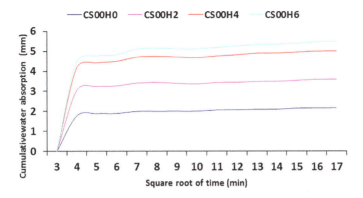

FIGURE 6.9 Sorptivity at 600°C.

is shown by the CS00H2 at 23°C. The sorptivity is reduced by 8.16 % comparing to normal concrete at 23°C.

The lowest value of sorptivity is shown by the CS00H2 at 400°C. The sorptivity is reduced by 50.26% comparing to normal concrete at 400°C. As the temperature increases, the sorptivity also increased. The sorptivity increased by 25.2 % and 27.61% for CS00H0, CS00H2 at 400°C comparing to 23°C. The lowest value of sorptivity is shown by the CS00H6 at 600°C. The CS00H6 sorptivity is reduced by 78.3% comparing to CS00H0 at 600°C. As the percentage of HDPE fibre increases, the sorptivity also decreased. The sorptivity is increased by 34.4% and 35% for CS00H2 and CS00H4 respectively at 600°C comparing to CS00H0. The sorptivity is lowest at 600°C comparing to other temperatures in all variations CS00H0, CS00H2, CS00H4, and CS00H6, and the highest sorptivity shown at 400°C comparing to other temperatures all variations CS00H0, CS00H2, CS00H4, and CS00H6.

6.6.2 Water Absorption After Applying Elevated Temperature

The water absorption at different elevated temperatures is shown in Figure 6.10. The specimens were prepared by cutting 100mm × 100mm × 100mm cubes in four equal parts i.e., 50mm × 50mm × 100mm after a water-curing period of 28 days. This test is done as per code ASTM C 642. The specimens of CS00 were tested after 28 days curing. It is observed that the water absorption is lowest at 0.2% HDPE fibre concrete and increasing with increase in percentage of HDPE fibre. The water absorption of CSS00H2 is decreased by 11.05% compared to CS00H0.

After the elevated temperature procedure followed, at 23°C the water absorption continuously increases with the increase in HDPE fibre content, whereas at normal procedure water absorption is reduced at CS00H2 compared to normal concrete. At the elevated temperature the water absorption is increased, and it is decreased with the increase in percentage of HDPE fibre in concrete at the elevated temperature. Highest water absorption is shown at 600°C. The lowest water absorption shown by CS00H6 comparing other percentages of HDPE-FRC.

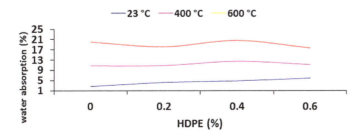

FIGURE 6.10 Water absorption of CS00H0 at different temperatures.

Beyond 0.2% HDPE fibre in concrete a network of fibre is formed, which probably reduced the flow of concrete. The reduction in compressive strength of HDPE fibre reinforced concrete has occurred probably due to low modulus of elasticity of HDPE fibre at higher percentages. The increment in flexural strength is due to the good bond between the fibre and the copper slag in concrete. The reduction in compressive strength at elevated temperatures is due to separation and spalling of concrete at higher temperatures. There is a rapid increase initially in cumulative water absorption and water absorption at elevated temperatures, probably due to creation of path between voids in concrete.

6.7 Conclusions

It was established that the main merits of HDPE fibres in the concrete mix improved the serviceability attributes. The following are the precise foremost conclusions:

a. The tensile strength and the flexural strength of concrete improved by the addition of HDPE fibres.

b. The compressive strength of concrete is not improved by the inclusion of HDPE fibres at 28 days but increased after 90 days water curing. The maximum workability is obtained at CS00H2 and at normal procedure water absorption is reduced at CS00H2 by 11%.

c. The optimum compressive strength occurred at CS00H4 and at CS60H2, increased by 30.39% and 14.20% comparing to normal concrete.

d. The optimum split tensile strength occurred at CS00H2 and at CS60H2, increased by 12.5% and 22.18% comparing to normal concrete.

e. Ductility is one of the main meritorious attributes of HDPE-FRC comparing the plain concrete and can keep a constant ductile capacity of concrete after cracking. The flexural strength is increased greatly at CS60H6 comparing to CS00H6.

f. The loss of weight is increasing with increase in temperature. The maximum loss of weight occurred with CS00H4.

g. The application of high temperature on the HDPE-FRC resulted in colour change, spalling, cracks, then meshy cracks, and also the strength was reduced.

h. HDPE fibres lowered water absorption of concrete by a distinguishable quantity. This indicates that HDPE-FRC is longer-lasting than the plain concrete. The water absorption of 0.2% HDPE-FRC is decreased by 11.05% compared to 0% HDPE fibre concrete at elevated temperatures.

i. Inclusion of HDPE fibres in concrete minimised the permeability of water by a recognisable quantity when measuring the depth of penetration comparing to the plain concrete, but at elevated temperature the sorptivity is increased with an increase in HDPE fibre and it is reduced at CS00H4. The lowest value of sorptivity is shown by the CS00H2 (0.2% HDPE-FRC) at 23°C to 400°C, and CS00H6 (0.6% HDPE-FRC) at 600°C.

REFERENCES

Al-Jabri, K. S. et al. (2006) Effect of copper slag and cement by-pass dust addition on mechanical properties of concrete', *Construction and Building Materials*, 20(5), pp. 322–331. doi: 10.1016/j.conbuildmat.2005.01.020.

Arino, A. M. and Mobasher, B. (1999) 'Effect of ground copper slag on strength and toughness of cementitious mixes', *Materials Journal*, 96(1), pp. 68–73. Available at: https://www.concrete.org/publications/internationalconcreteabstractsportal.aspx?m=details&ID=430&m=details&ID=430.

Brindha, D. and Nagan, S. (2011) 'Durability studies on copper slag admixed concrete', *Asian Journal of Civil Engineering*, 12(5), pp. 563–578.

EN 1992-1-2 (2004) Eurocode 2: Design of concrete structures - Part 1-2: General rules - Structural fire design. EUROPEAN C.

Husem, M. (2006) 'The effects of high temperature on compressive and flexural strengths of ordinary and high-performance concrete', *Fire Safety Journal*, 41(2), pp. 155–163. doi: 10.1016/j.firesaf.2005.12.002.

Kalifa, P., Chéné, G. and Gallé, C. (2001) 'High-temperature behaviour of HPC with polypropylene fibres', *Cement and Concrete Research*, 31(10), pp. 1487–1499. doi: 10.1016/s0008-8846(01)00596-8.

Kalifa, P., Menneteau, F. D. and Quenard, D. (2000) 'Spalling and pore pressure in HPC at high temperatures', *Cement and Concrete Research*, 30(12), pp. 1915–1927. doi: 10.1016/S0008-8846(00)00384-7.

Isabel Sanchez, M., de Rojas, M.I.S., Rivera, J., Frías, M. and Marín, F. (2008) 'Review Use of recycled copper slag for blended cements', *Journal of Chemical Technology & Biotechnology*, 83(May), pp. 1163–1169. doi: 10.1002/jctb.

Najimi, M. and Pourkhorshidi, A. R. (2011) 'Properties of concrete containing copper slag waste', *Magazine of Concrete Research*, 63(8), pp. 605–615. doi: 10.1680/macr.2011.63.8.605.

Peng, G. F. et al. (2006) 'Explosive spalling and residual mechanical properties of fiber-toughened high-performance concrete subjected to high temperatures', *Cement and Concrete Research*, 36(4), pp. 723–727. doi: 10.1016/j.cemconres.2005.12.014.

Pešić, N. et al. (2016) 'Mechanical properties of concrete reinforced with recycled HDPE plastic fibres', *Construction and Building Materials*, 115, pp. 362–370. doi: 10.1016/j.conbuildmat.2016.04.050.

Pliya, P., Beaucour, A. L. and Noumowé, A. (2011) 'Contribution of cocktail of polypropylene and steel fibres in improving the behaviour of high strength concrete subjected to high temperature', *Construction and Building Materials*. Elsevier Ltd, 25(4), pp. 1926–1934. doi: 10.1016/j.conbuildmat.2010.11.064.

Poon, C. S., Shui, Z. H. and Lam, L. (2004) 'Compressive behavior of fiber reinforced high-performance concrete subjected to elevated temperatures', *Cement and Concrete Research*, 34(12), pp. 2215–2222. doi: 10.1016/j.cemconres.2004.02.011.

Rao, M. M. (2018) 'Investigation on Properties of PET and HDPE Waste Plastic Concrete', *International Journal for Research in Applied Science and Engineering Technology*, 6(3), pp. 495–505. doi: 10.22214/ijraset.2018.3080.

Shi, C., Meyer, C. and Behnood, A. (2008) 'Utilization of copper slag in cement and concrete', *Resources, Conservation and Recycling*, 52(10), pp. 1115–1120. doi: 10.1016/j.resconrec.2008.06.008.

Tanyildizi, H. (2009) 'Statistical analysis for mechanical properties of polypropylene fiber reinforced lightweight concrete containing silica fume exposed to high temperature', *Materials and Design*. Elsevier Ltd, 30(8), pp. 3252–3258. doi: 10.1016/j.matdes.2008.11.032.

Kodur, V.K.R. (2004) 'Spalling in High Strength Concrete Exposed to Fire – Concerns, Causes, Critical Parameters and Cures', *ASCE Structures Congress* 2000, pp. 1–9. Available at: https://ascelibrary.org/doi/10.1061/40492%282000%29180.

Zheng, W., Li, H. and Wang, Y. (2012) 'Compressive behaviour of hybrid fiber-reinforced reactive powder concrete after high temperature', *Materials and Design*. Elsevier Ltd, 41, pp. 403–409. doi: 10.1016/j.matdes.2012.05.026.

7

Utilization of Geo-Waste in Production of Geo-Fiber Papercrete Bricks

Areej Palekar, Suraj Patil, and Uma Kale
M.H. Saboo Siddik College of Engineering, Mumbai, India

CONTENTS

7.1 Introduction 53
    7.1.1 Objectives 54
7.2 Methodology 54
    7.2.1 Materials 54
7.3 Experimental Procedure 55
    7.3.1 Preparation of Paper Pulp 55
    7.3.2 Fabrication of the Mold 55
    7.3.3 Casting of Papercrete Cubes and Bricks 55
7.4 Testing and Results 57
7.5 Discussion 59
7.6 Conclusion 59
Acknowledgement 60
References 60
IS Codes Referred 60

7.1 Introduction

Due to the increasing population and their respective demands, creation of solid waste has become a major sign of concern for the upcoming youth and the disposal of solid waste is the major issue. India generates an estimated 0.143 million tonnes of solid waste per day as per CPCB (Central Pollution Control Board, 2016). An international survey conducted by the EPA (Environment Protection Authority) has found out that every year 4 million tonnes of paper is produced every year. So, one can imagine how much paper waste is generated. From the efficiency of recycling the wastepaper, we have found that only 70–80% wastepaper can be recycled, and the remainder is brought for landfilling. Paper waste being one of the major components of geo-solid waste is the third in the list after food waste and wood waste. Whereas the fly ash waste from the boilers, due to its uselessness in most applications, ends up in landfills where it poses serious hazards to the environment and human health.

The crisis of building material shortfalls is faced by the relevant construction industry to satisfy the raised demand for building materials. It has become a threat and a challenge to the engineers, scientists, and researchers to seek out another solution and convert the geo-solid waste into some helpful construction material. Consequently, this study is meted out so as to resolve these varieties of problems faced. Papercrete is a composite material created by the victimization of wastepaper, fly ash, coconut fiber, thereby it reduces the volume of cement within the concrete, thereby reducing the cost of building material, once combined and cured, these materials manufacture a product such as concrete; however it is light-weight (Shermale & Verma, 2017). By adding geo-fiber waste in concrete, we can make our structure an economically sound and profitable substitute to landfill and

DOI: 10.1201/9781003217619-8

incinerator. It was decided to study papercrete blocks and bricks to measure its strength, workability, and other properties to determine whether it is suitable to be used as a construction material. Because of its light weight in nature, it can be used for interior walls in high-rise buildings as well as it can be used for manufacturing the concrete bricks/blocks and non-structural members that could reduce the dead load of the structure. There is no specific code provision for the mix design of papercrete as it is still in its developing stage. (Myriam Marie, 2017). This material will help in meeting the sustainable development requirement.

7.1.1 Objectives

1. By using geo-solid waste consisting of paper waste, fly ash and coir we can reduce the quantity of solid waste generation to some extent.
2. The purpose behind this geo-fiber papercrete brick is to fulfill the requirements of building material to replace the costly and tedious and time-consuming process of brick manufacturing.
3. By manufacturing these eco-friendly bricks, we contribute to the modern era of green construction.

7.2 Methodology

The methodology consists of the following steps:

1. Raw materials collection
2. Preparation of paper pulp by geo-solid waste
3. Mold Fabrication
4. Preparation of the mix
5. Testing of papercrete bricks and cubes.
6. Results

7.2.1 Materials

Materials and their collection is the first and very important thing in the execution of the project. In this experiment geo-waste products like paper, fly ash, coconut coir, are used as the main component along with cement, sand and water. The materials which had been used for the production of geo-fiber paper-crete bricks are specified below.

A. Cement
   Cement is the main material in the construction industry for any type of construction work. Cement is used as the binding material in geo-fiber papercrete bricks manufacturing. We have used 53 Grade Ordinary Portland Cement that fulfills the recommendation of IS: 8112-1989. The characteristics properties of OPC are as follows
   1. Specific Gravity is 3.15
   2. Fineness is 412.5 m^2/kg
   3. Initial setting time is 75 minutes
   4. Final setting time is 270 minutes
B. Paper
   Paper is the main component of geo-fiber papercrete. Paper is a natural polymer consisting of wood cellulose. This wood cellulose is made up of units of the monomer of glucose. These celluloses are water-insoluble. The papers which are used in the experiment are newspapers, waste papers from dustbins, waste paper from paper recycling mills. Coating the cellulose fiber of paper with Portland cement encased the strength of geo-fiber papercrete mixture by forming a cement matrix.

C. Water

Water is an important component of the geo-fiber papercrete as it is used in a chemical reaction with the cement. Potable water is used in the soaking and mixing operation. The water should be added in proportion to get the required mix. Water which is used should have an almost pH value between 6–8 and should not contain any soluble impurities. The pH observed was 6.9 while performing the experiment.

D. Sand

Fine aggregates river sand and Manufactured sand (M-Sand) passing 4.75mm IS sieve as per the specifications in IS: 383-1970 were used. The sand used was clean, sharp, heavy, and gritty to touch. It should be free from clay, mica and other impurities to give a good mix. Sand should be perfectly dry before it is used.

E. Fly ash

Fly ash is produced by small to medium factories such as paper mills when they incinerate materials to produce energy. Due to its uselessness, it ends up in landfills, which in turn pollutes the soil quality. So to make use of this, we have used P-100 fly ash in our experiment, which results in an increase of the strength of papercrete and reduction in the volume of cement.

F. Coconut Fiber

Coconut fiber is obtained by de-husking between skin and shell. These are multi-cellular, ligno-cellulosic, hard, a very coarse, and a rigid variety of natural fruit fiber. Its advantages are agro-renewability, biodegradability and a good blend of strength, length, extensibility, moisture regain and high durability and resistance against sunlight, saline water, and microbes. Coconut fiber will act as reinforcement in the geo-fiber papercrete. (Anokha Shilin, 2016)

7.3 Experimental Procedure

The experimental procedure for manufacturing of papercrete cubes and bricks is as follows.

7.3.1 Preparation of Paper Pulp

To use the paper in the concrete mix, it should be weighted first as it will be difficult afterward to calculate the original weight of paper. Take a water bucket fill it 2/3 with water. Before immersion of wastepaper into water the papers are likely to be shredded into small pieces, this shredded paper should be immersed properly in such a way that each piece is wetted. Place the bucket aside undisturbed for 2–3 days. Once this period is over remove the paper from water and put in the machine mixer to get the homogeneous paper pulp. The extra water should be drained off. (Myriam Marie, 2017)

7.3.2 Fabrication of the Mold

Before preparing the papercrete mix, first, we have to make a papercrete cube mold of size 150mm × 150mm × 150 mm. While fabricating, the corners of the mold should be watertight to avoid the concrete leakage and the interior surface of the assembled mold was coated with mold oil to prevent adhesion of concrete. We have used steel mold for papercrete cubes whereas, for the fabrication of brick mold, plywood was used to make a mold of size 190mm × 90mm × 90mm.

7.3.3 Casting of Papercrete Cubes and Bricks

After the production of paper pulp, the pulp is ready to mix with the other ingredient which include cement, sand, fly ash, coconut fiber and water in proportion. From the research up till now there is no specific procedure for casting the bricks and the procedure followed in this investigation was as per our conveniences and there is no hard mix of papercrete found yet. So we have performed this experiment based on our self-trial mixes. Some of the trail mix which we have adopted are given in Table 7.1.

TABLE 7.1

Different Mix Proportions for Papercrete Components

Trial Mix no.	Component	Trial Mix	Paper (gms)	Cement (gms)	Sand (gms)	Fly Ash (gms)	Coconut Coir (gms)
1	Bricks	paper + cement + sand	1000	500	500		
2	Bricks and blocks	paper + cement + sand + fly ash	1000	500	500	500	
3	Bricks, blocks, RCC grills	paper +cement + sand + fly ash + coconut coir	1000	500	500	500	500

FIGURE 7.1 Papercrete brick specimen.

The ingredients taken were on a weight batching method as use of weight system in batching facilitates accuracy, flexibility and simplicity. (Shetty, n.d.)

After the formation of the papercrete mix, it should be filled in cubes and bricks mold within 30 minutes. The compaction of papercrete is done by vibratory compaction machine.

- For Brick Specimen-
- After the compaction, the brick molds are kept undisturbed for the 24 hours as shown in Figure 7.1. After a period of 24–30 hrs, the bricks should be removed from the mold, and should be kept in sun for drying at least for 14 days.
- For Cube Specimen-

Three cubes were casted for each trial mix. The cubes specimens are kept in mold for 24 hours and after 24 hours the cubes were removed from the mold for curing and marked as shown in Figure 7.2. The curing period was 28 days.

FIGURE 7.2 Papercrete cube specimen.

Utilization of Geo-Waste 57

7.4 Testing and Results

To determine the efficiency, usefulness and economic soundness of the papercrete bricks and cubes casted testing is done.

Testing and Results include:

1. Compressive Strength Test

   Compressive strength is the quiet one of the most important testing done on cubes and bricks. The compressive strength of 3 cubes and bricks of each trial mix were measured at 28 days and 14 days respectively and an average compressive strength was taken into consideration.

   - For Papercrete Bricks

     To test the papercrete bricks the specimen was kept under a universal testing machine as per IS: 3495 (Part 1)-1992. For uniform distribution of load, the bricks were placed between two plywood sheets and the load was applied at the rate of 170 kg/cm^2 per minute. The compressive strength of the papercrete varies with composition refer to Table 7.2 for the observed result after 14 days. It was observed while testing the specimens that the bricks did not crush or completely collapse, it just compressed like squeezing a rubber. It was evaluated that these bricks have elastic behavior and are less brittle as compared to conventional common burnt clay bricks.

     1. Average compressive strength for Mix 1 after 14 days is 4.65 N/mm^2
     2. Average compressive strength for Mix 2 after 14 days is 1.7 N/mm^2
     3. Average compressive strength for Mix 3 after 14 days is 3.43 N/mm^2

   - For Papercrete Cubes

     At the age of 28 day, the papercrete cube specimens were tested under the compressive testing machine as per IS: 516-1959.The load shall be applied without any shock and increased continuously at a rate of approximately 140kg/sg/min until the resistance of the specimen to the increasing load breaks down and no longer load can be sustained by cube specimen. Table 7.3 shows the compressive strength result.

     1. Average compressive strength for Mix 1 after 28 days is 3.2 MPa
     2. Average compressive strength for Mix 2 after 28 days is 2.7 MPa
     3. Average compressive strength for Mix 3 after 28 days is 4.7 MPa

   Other tests were conducted on brick specimen only which are as follows.

2. Weight Test

   As we have found out that mixing the paper pulp with cement, sand, crushed stones along with coconut fiber and fly ash, the weight will be decreased. After drying we have weighted these

TABLE 7.2

Compressive Strength of Bricks after 14 Days of Sun Drying

Trial Mix Number	14 Days Compressive Strength in N/mm^2
1	7.2
	3.48
	3.28
2	1.72
	1.67
	1.71
3	3.58
	3.24
	3.48

TABLE 7.3

Compressive Strength of Cubes after 28 Days Testing

Trial Mix Number	28 Days Compressive Strength in MPa
1	3.2
	3.4
	3
2	2.6
	3.2
	2.3
3	4.4
	4.5
	5.2

papercrete bricks and compare it with conventional clay bricks. The weight of conventional brick was coming out to be 2.26 kg whereas the weight of papercrete brick observed was 1.5 kg. We have found out that the weight of papercrete bricks is lighter than the conventional common burnt clay bricks.

3. Water Absorption Test

    For the water absorption test first, the bricks were kept in an oven at 100 degrees Celsius for the time when constant weight is observed to get the dry weight. After that, the bricks were immersed in water for 24hrs and then weighed again, and the difference in weight indicates the amount of water absorbed by the brick. The water absorption was found to be 21%.The test was conducted as per IS 3495(Part 2):1992. It is used to determine whether the bricks are suitable for water-logged areas (Manoj Kumar, 2017).

4. Fire Test

    Fire test is also the most important property of the material as a purpose of safety. As paper is an organic material, it easily catches fire in an open flame. So the test was conducted on the papercrete bricks in an open flame for 30 minutes and results have been shown that these bricks cannot easily catch fire in presence of an open flame at least till 10 min, but when it is fired for a longer time it would burn and get the ash form. Interior plaster and exterior stucco should be provided on these bricks, to prevent them from getting burnt.

5. Structure Test

    This is the basic property of a material that should be homogeneous. For this, we have cut the bricks with the saw blade and observed the brick very carefully. We found that all the particles of the papercrete mix are properly compacted which in turn showed that it has a homogeneous structure.

6. Cutting and Gluing Test

    As we are aware of construction sites where the laborers are unable to cut the conventional bricks in the desired shape required and hence lots of wastage of bricks occur. But in the case of papercrete bricks, we have found out that we can easily cut the bricks in the desired form and we also can get it glued as per the requirement.

7. Soundness Test

    Two bricks were struck with each other to determine the soundness property of bricks. It produced a clear ringing sound which indicates good quality of brick.

8. Efflorescence Test

    For the presence of minerals and salts, the bricks were immersed in water for 2hrs and after that, were sun-dried. From the observation after drying, we found out that there was no white or grey mark/spots on the bricks. The liability to efflorescence was reported as nil as per IS 3495 (Part 3):1992. It indicates that the bricks were free from soluble salts and alkaline matter.

Utilization of Geo-Waste

FIGURE 7.3 RCC Grill.

7.5 Discussion

The following discussion is considered after observing the results.

1. The compressive strength test is one of the important testing parameters from the compressive strength test. We found out that the geo-fiber papercrete cube is giving the best compressive strength among the 3 trial mixes performed in the experiment.
2. The various components can be manufactured by taking the combination of material in proportion such as:
   - Paper pulp and cement to make decorative item
   - Paper pulp + cement + sand = bricks
   - Paper pulp + cement + and + fly ash + coconut coir = bricks, blocks, Structural components like RCC grills and.RCC grills were prepared based on the above mix as shown in Figure 7.3 and can be used as the windows for the staircase room.

7.6 Conclusion

1. From the water absorption test, we found out that water absorption value exceeds 20% i.e., recommended by IS 3495 (Part 2):1992, which makes this geo-fiber papercrete bricks not suitable to be used in the exterior walls.
2. Due to water absorption value exceeding the limit as per IS: 3495(Part 2)-1992, therefore, it can be used by applying the coat of geo-bond or latex or some other proofing materials.
3. Due to the homogeneous structure and flexibility of geo-fiber papercrete bricks, the use of these bricks is of greater advantage in earthquake-prone zones.
4. It has been found out that after mixing the ingredients of geo-solid waste along with concrete, the dead weight of the brick is reduced by 2/3 of the weight of the conventional brick therefore, there is an additional advantage as an economic point of view and also it contributes to the modern trend of green and eco-friendly construction.
5. As this geo-fiber papercrete brick is made up of wastepaper, so, on-site, we can manually cut the brick into required dimensions with the saw blade, thus we can reduce the loss of bricks as would happen in the case of conventional fired clay bricks.
6. These bricks are majorly made-up of waste material, which are of no use other than that of landfilling. Thus we can approach the purpose of green construction and solid waste management.
7. These bricks can be used for the construction of non-structural walls in building.

ACKNOWLEDGEMENT

We would like to extend our vote of thanks to IconSWM for providing us with a golden opportunity to represent our work. Later we are thankful to Civil Engineering Department of M.H.Saboo Siddik College of Engineering, Byculla, Mumbai for providing us the instruments and laboratory to work.

REFERENCES

Anokha Shilin, Feba Halus, Mariyam Habeeba, M.K. Merin, Maria Johny, 2016. Papercrete. *International Journal of Engineering Research and Technology*, 5(13), ISSN: 2278-0181, ICCEECON - 2K16 Conference Proceedings.

Manoj Kumar, M., Uma Maheshwari, G., 2017. Papercrete. *International Journal of Science, Engineering and Technology Research,* 7(8), pp. 1289–1297, ISSN: 2278–7798.

M.S. Shetty, n.d. *Concrete Technology*. s.l.:S. Chand & Company, pp. 238–240.

Myriam Marie Delcasse, R.V., 2017. Papercrete bricks-An alternative sustainable building material. *International Journal of Engineering and Research Application,* 7(3), pp. 09–14.

Yogesh Shermale, Mahaveer B. Verma, 2017. Properties of papercrete concrete: Building Material. *IOSR Journal of Mechanical and Civil engineering,* 14(2), pp. 27–32, ISSN: 2320-334X.

IS CODES REFERRED

1. IS: 3492 (Part 1–4)-1992, Methods of Tests of burnt clay building bricks.
2. IS: 516-1959, Methods of test for strength of concrete.
3. IS: 383-1970, Specification for coarse and fine aggregates from natural sources for concrete.

8

Effect of Slag Addition on Compressive Strength and Microstructural Features of Fly Ash Based Geopolymer

Dipankar Das
Tripura University, India

Alok Prasad Das
Rama Devi Women's University, India

Prasanta Kumar Rout
Tripura University, India

CONTENTS

8.1 Introduction 61
8.2 Materials and Methods 62
    8.2.1 Materials 62
    8.2.2 Methods 63
8.3 Results and Discussion 63
8.4 Conclusions 66
Acknowledgements 67
References 67

8.1 Introduction

Geopolymeric materials are one of the alternative binding materials known after the Ordinary Portland Cement (OPC). Geopolymers are amorphous, three-dimensional short-range order inorganic polymeric material which can be formed by the reaction between an aluminosilicate source material and an alkaline solution having a high pH value (Das & Rout, 2019). Based on the geopolymer structure, geopolymers are categorized into three basic systems, such as poly (sialates) (-Si-O-Al-O-), poly(sialate-siloxo) (-Si-O-Al-O-Si-O-) and poly(sialate-disiloxo) (-Si-O-Si-O-Al-O-Si-O-) (Yun-Ming et al., 2016; Kai, Zhang & Liew, 2020). The general empirical formula for geopolymer is (Nath et al., 2016; Marcin, Sisol & Brezani, 2016; Das & Rout, 2021a):

$$Mn\left[-\left(SiO_2\right)z - AlO_2\right]n.wH_2O$$

(Where M represents an alkaline element, symbol (-) indicates the presence of bond, z is 1, 2, 3 or higher up to 32 and n is the degree of polymerization). The mechanism of geopolymerization consists of three steps, i.e., (i) dissolution of the aluminosilicate source material in the alkaline solution, (ii) transportation or orientation, (iii) polycondensation (Xu & Van Deventer, 2000). These geopolymeric binders can reduce the greenhouse gas (CO_2) emission by 80%, which was entailed by the conventional cement industry (Duxson et al., 2007). Due to their excellent mechanical properties, better thermal and chemical

DOI: 10.1201/9781003217619-9

resistance, more durable, low heating temperature and thus low energy consumption, the use of geopolymeric materials in construction is gaining importance (Singh et al., 2015).

Fly ash (FA) is an industrial by-product generated in the process of burning pulverized coal in electrical power generating plants (Das & Rout, 2021c). Due to the amorphous silica and alumina, along with sub-micron particle size and shapes, this makes fly ash powder favorable or suitable for making geopolymeric material (Toniolo & Boccaccini, 2017). The final geopolymeric product produced by the alkali activation of FA is having good strength, low shrinkage and good thermal resistivity. It has been reported that the main reaction product in the fly ash based geopolymer is N-A-S-H (N-Na_2O, A-Al_2O_3, S-SiO_2 and H-H_2O) gel, as observed in SEM (Marjanovi et al., 2014; Das & Rout 2021b). In natural curing at room temperature, the FA based geopolymer shows low reactivity, which results in low setting time and low strength gain (Winnefeld et al., 2010; Palomo, 2007; Kumar, Kumar & Mehrotra, 2010). To avoid this issue and to get better strength, the fly ash particles are processed by mechanical activation (Marjanovi et al., 2014) and also by the addition of ground granulated blast furnace slag (GGBS) (Kumar, Kumar & Mehrotra, 2010). The term mechanical activation can be defined as the process affected by mechanical energy, which improves the chemical reactivity of the solids through physio-chemical changes without altering the chemical composition of the raw material. There are various types of mechanical activation, such as a) mechanical dispersion, b) surface activation, c) mechano-chemical (structural) activation (Mucsi, 2016). Kumar et al. studied the mechanical activation of fly ash, where it has been reported that mechanical activation in eccentric vibratory mill increases the reactivity of fly ash (Kumar & Kumar, 2011). GGBS is another industrial by-product obtained by quenching molten iron slag from a blast furnace in steel plants. This granular product is dried and ground into a fine powder by using a ball mill. The main product formed by the reaction between slag and alkaline activator is calcium silicate hydrate along with the aluminum (Al) in the structure, which is also known as C-A-S-H (C- CaO, A- Al_2O_3, S-SiO_2 and H-H_2O) gel (Lee & Lee, 2015). The main limitation of alkali-activated slag-based geopolymer is poor workability and fast setting time. To get rid of this situation, there could be a possibility to synthesized geopolymer by blending both FA and GGBS powder by activating with alkaline activators. This leads to the formation of binding gels of Na_2O-CaO-Al_2O_3-SiO_2 (Ismail, Bernal et al., 2014a, b). Puertas et al. (2000) studied the influence of high-calcium slag on fly ash based geopolymer, where it has been seen that GGBS reacts almost completely, and the compressive strength increases with an increase in GGBS. Gong et al. (2014) also reported that the inclusion of GGBS increases the compressive strength of the fly ash and GGBS based geopolymer. So, the objective of this present investigation is to study the effect of GGBS powder particles on compressive strength and microstructural features of FA based geopolymer. The observed strength value of the final geopolymer has been explained by the help of the SEM fractographs.

8.2 Materials and Methods

8.2.1 Materials

The raw materials used in this study were class F fly ash (FA), which was collected from National Thermal Power Corporation Limited (NTPC), Bongaigaon, Assam, India and Ground Granulated blast furnace slag (GGBS), an industrial by-product from iron production plants. The GGBS is received initially in a granulated form and then milled into a fine powder with a conventional ball mill. Sodium hydroxide (NaOH) (assay–98.0%; make: Loba Chemie Private Limited), Sodium silicate solutions or water glass (Na_2SiO_3) (Assay – Na_2O: 7.5–8.5%, SiO_2: 25–28%, make: Loba Chemie Private Limited) were used as an alkaline activator. The average particle size of FA and GGBS powder was measured by using a Malvern Mastersizer particle size analyzer, UK. The chemical composition of the powder was measured using X-ray fluorescence (XRF). The morphology of both powders was characterized by field emission scanning electron microscopy (Model: FEI Quanta FEG 200, Make: ThermoFisher SCIENTIFIC and Model: Sigma-300, make: Carl Zeiss).

Effect of Slag Addition 63

TABLE 8.1

Mix Design of Geopolymer Paste

Sl. No.	Sample ID	Concentration (NaOH) Solution	Na$_2$SiO$_3$/NaOH Ratio	Starting Powder		Alkaline Solution		Liquid/Solid Ratio	Curing Condition
				FA (g)	GGBS (g)	NaOH Solution (g)	Na$_2$SiO$_3$ Solution (g)		
1	FA0-S100	14	1	0	100	15	15	0.3	60°C for
2	FA25-S75	14	1	25	75	15	15	0.3	24 hours
3	FA50-S50	14	1	50	50	15	15	0.3	
4	FA75-S25	14	1	75	25	15	15	0.3	
5	FA100-S0	14	1	100	0	15	15	0.3	

8.2.2 Methods

The alkaline solutions having a concentration 14 M was prepared by dissolving NaOH pellets in distilled water. Initially, the solution gets heated due to the exothermic reaction during the dissolution process, so the solutions are stored at room temperature for 24 hours before use (Kumar, Kumar & Mehrotra, 2010). FA and GGBS powder were mixed with 14 M NaOH solution and stirred manually for two minutes. Then the appropriate amounts of Na$_2$SiO$_3$ solutions were added and mixed manually for another two minutes to get a homogeneous paste. The pastes were casted in a cylindrical plastic mold having dimension 5cm x 4cm (length x diameter) and vibrated for few minutes to remove the entrapped air bubbles. After 24 hours, the samples were de-molded, followed by artificial curing in an oven at 60°C for 24 hours. Table 8.1 shows the mix designs of the geopolymer paste and in the table, FA indicates fly ash and S indicates ground granulated blast furnace slag with appropriate ratio.

The compressive strength test was performed on a digital compression testing machine (Model: AIM-314E-DG-1, make: Aimil Limited). Three samples from each group were tested and their mean values were reported in this study. The fracture surface of geopolymer samples was observed by using field emission scanning electron microscopy (Model: Sigma-300, make: Carl Zeiss).

8.3 Results and Discussion

Table 8.2 shows the chemical composition of FA and GGBS powder. The fly ash powder is having a substantial amount of SiO$_2$ and Al$_2$O$_3$ along with some minor oxide components such as Fe$_2$O$_3$, TiO$_2$, CaO, MgO, ZnO, Na$_2$O, SrO, SO$_3$ and K$_2$O, etc. and GGBS powder has SiO$_2$, Al$_2$O$_3$ and CaO as a major oxide component along with some minor oxide components such as Fe$_2$O$_3$, MgO, and SO$_3$. The LOI (loss on ignition) of FA and GGBS powder is 0.47 and 1.75, respectively.

Figure 8.1 shows the particle size distribution (PSD) curves and Table 8.3 listed the characteristic particle diameters (μm) for both the powder. Figure 8.2 shows the XRD pattern of FA and GGBS powders. Alpha quartz and mullite are the major crystalline phases present in FA powder. Furthermore, a broad hump has been seen between two theta values, ranging from 18° to 35°, indicates the presence of

TABLE 8.2

Oxide Composition (wt. %) of FA and GGBS Powder

Oxides	SiO$_2$	Al$_2$O$_3$	CaO	Fe$_2$O$_3$	TiO$_2$	MgO	K$_2$O	Na$_2$O	MnO	SrO	ZnO	SO$_3$	LOI
FA powder	55.6	29.80	1.59	5.91	1.63	1.08	1.94	0.23	0.05	0.04	0.03	0.45	0.47
GGBS powder	33.42	16.81	37.2	0.97	-----	8.38	-----	-----	-----	-----	-----	0.15	1.75

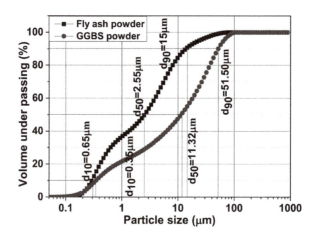

FIGURE 8.1 Particle size distribution (PSD) of FA and GGBS powder.

TABLE 8.3

Characteristic Particle Diameters (μm) of Fly Ash and GGBS Powder

Description	Characteristic Particle Diameters (μm)		
Particle Diameters (μm)	d_{90}	d_{50}	d_{10}
FA powder	15 μm	2.55 μm	0.65 μm
GGBS powder	51.50 μm	11.32 μm	0.35 μm

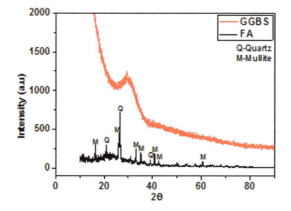

FIGURE 8.2 XRD graphs of FA and GGBS powder.

an aluminosilicate amorphous phase. But, unlike FA powder, the GBFS powder shows a board hump in between 25° to 35° attributed to the presence of a glassy or amorphous phase and does not show any peak for crystalline phases (Ismail, Bernal et al., 2014a, b).

Figure 8.3 shows the morphology of FA and GGBS powders. Figure 8.3a, SEM micrograph of FA powder shows mostly round and spherical shapes with the smooth texture of different particle sizes. Further, it exhibits a very little amount of angular and irregular shaped particles with a conglomeration of spherical particles. Figure 8.3b, an SEM micrograph, shows mostly irregular flakes, anomalous shapes with angular and sharp edges particles look like broken glasses.

The effect of GGBS addition on the compressive strength of geopolymeric specimens under artificial curing is shown in Figure 8.4. From this figure, it can be seen that the sample ID FA100-S0 has very

Effect of Slag Addition 65

FIGURE 8.3 Morphology of starting raw materials (a) FA powder and (b) GGBS powder.

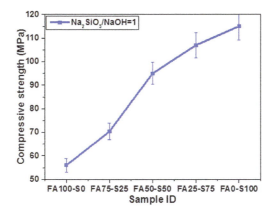

FIGURE 8.4 Effect of GGBS addition on compressive strength of geopolymeric specimen under artificial curing.

low strength as compared to the other sample. The main reaction product for fly ash based geopolymer is N-A-S-H gel or sodium aluminosilicate gel and the main reaction product for ground granulated blast furnace slag based geopolymer is C-A-S-H gel. The addition of ground granulated blast furnace slag in fly ash based geopolymer improved the compressive strength that is due to the C-A-S-H gel formation (Kumar, Kumar & Mehrotra, 2010; Buchwald, Hilbig & Kaps, 2007; Buchwald, Tatarin & Stephan, 2009; Yao et al., 2009). The presence of soluble calcium content in the GGBS also has a direct effect on the compressive strength (Shadnia & Zhang, 2017). Ismail et al. reported that the main reaction product for the fly ash and ground granulated blast furnace slag based geopolymer is the presence of N-A-S-H and C-A-S-H gels or hybrid C-N-A-S-H gel (Ismail, Bernal et al., 2014a, b). Curing temperature and curing time also play vital roles in increasing the compressive strengths of geopolymeric specimens. Table 8.4 shows the compressive strength data of the geopolymer sample prepared with various proportions of fly ash and ground granulated blast furnace slag.

Figure 8.5 (a–e) shows the few representative SEM fractographs of the geopolymeric samples ID FA100-S0, FA75-S25, FA50-S50, FA25-S75 and FA0-S100. The microstructure of the GGBS-FA based geopolymeric specimen shows that most of the particles are fully reacted, and some particles are partially reacted. The microstructure of geopolymeric specimens also shows some micro-cracks along with some micropores. The sample ID FA25-S75 and FA0-S100 shows a very dense and compact structure due to the full reaction of starting raw materials and the sample ID FA100-S0, FA75-S25 and FA50-S50 shows less dense structure compared to the sample ID FA25-S75 and FA0-S100 due to the unreacted and the partial reaction of fly ash powder.

TABLE 8.4

Compressive Strength Data of the Geopolymer Sample

Sl No.	1	2	3	4	5
Sample ID	FA100-S0	FA75-S25	FA50-S50	FA25-S75	FA0-S100
Maximum compressive strength (MPa)	56	70	95	107	115

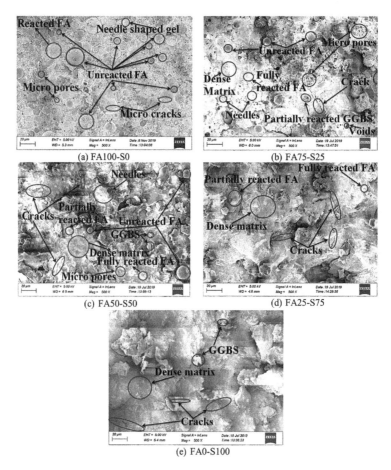

FIGURE 8.5 (a–e): Few representative SEM fractographs of geopolymeric specimens.

8.4 Conclusions

In this study, artificially cured alkali-activated GGBS-FA based geopolymer samples with varying percentages were prepared. The mechanical properties and microstructural features were studied, and the major conclusions are as follows.

The compressive strength was improved by the addition of GGBS into the FA geopolymer and reached a maximum compressive strength of 107 MPa for the sample ID FA25-S75 and 115 MPa for the sample ID FA0-S100. The fractograph of fly ash geopolymer contains unreacted fly ash particles, partially reacted fly ash particles, and pores embedded in a continuous alumina-silicate N-A-S-H gel denser matrix. An additional crystalline C-A-S-H gel was formed in conjunction with the amorphous aluminosilicate N-A-S-H gel, which modified the microstructure and thus associated the better compressive strength (Marjanovi et al., 2015).

Effect of Slag Addition

ACKNOWLEDGEMENTS

The authors are grateful to the GBPNIHESD (IERP funded), Government of India for supporting the project entitled "Production of Geopolymer based construction material from Fly ash: An industrial waste" We are very thankful to Mr. Ekonthung Ngullie (DGM) and the team (NTPC Limited, Bongaigaon, Assam, India) for providing us fly ash powder to carry out the research work. The authors are also very much thankful to the Central Instrumentation Centre, Tripura University, Agartala, India, for allowing us to avail the facilities of FESEM.

REFERENCES

Buchwald, A., Hilbig, H. and Kaps, C. (2007) 'Alkali-activated metakaolin-slag blends - Performance and structure in dependence of their composition', *Journal of Materials Science*, 42(9), pp. 3024–3032. doi: 10.1007/s10853-006-0525-6.

Buchwald, A., Tatarin, R. and Stephan, D. (2009) 'Reaction progress of alkaline-activated metakaolin-ground granulated blast furnace slag blends', *Journal of Materials Science*, 44(20), pp. 5609–5617. doi: 10.1007/s10853-009-3790-3.

Das, D. and Rout, P. K. (2019) 'Utilization of thermal industry waste: *From trash to cash Carbon – Science and Technology*', 11(2), pp. 43–48.

Das, D. and Rout, P. K. (2021a) 'Synthesis, Characterization and properties of fly ash based geopolymer materials', *Journal of Materials Engineering and performance*, 30(5), pp. 3213–3231. doi: 10.1007/s11665-021-05647-x

Das, D. and Rout, P. K. (2021b) 'Synthesis and characterization of fly ash and GBFS based geopolymer material', *Biointerface Research in Applied Chemistry*, 11(6), pp. 14506–14519. doi: 10.33263/BRIAC116.1450614519.

Das, D. and Rout, P. K. (2021c) 'Industrial solid wastes and their resources." In *Emerging Trends in Science and Technology*, edited by Nagendra Singh, Mukesh Kumar Kumawat, 1st Edition, 96–101. Bhumi Publishing.

Duxson, P. et al. (2007) 'The role of inorganic polymer technology in the development of "green concrete"', *Cement and Concrete Research*, 37(12), pp. 1590–1597. doi: 10.1016/j.cemconres.2007.08.018.

Gong, W. et al. (2014) 'Effect of blast furnace slag grades on fly ash based geopolymer waste forms', *Fuel*, 133, pp. 332–340. doi: 10.1016/j.fuel.2014.05.018.

Ismail, I., Bernal, S. A., et al. (2014a) 'Cement & Concrete Composites Modification of phase evolution in alkali-activated blast furnace slag by the incorporation of fly ash', 45, pp. 125–135.

Ismail, I., Bernal, S. A., et al. (2014b) 'Modification of phase evolution in alkali-activated blast furnace slag by the incorporation of fly ash', *Cement and Concrete Composites*, Elsevier Ltd, 45, pp. 125–135. doi: 10.1016/j.cemconcomp.2013.09.006.

Kai, M. F., Zhang, L. W. and Liew, K. M. (2020) 'Carbon nanotube-geopolymer nanocomposites: A molecular dynamics study of the influence of interfacial chemical bonding upon the structural and mechanical properties', *Carbon*, Elsevier Ltd, 161, pp. 772–783. doi: 10.1016/j.carbon.2020.02.014.

Kumar, S. and Kumar, R. (2011) 'Mechanical activation of fly ash: Effect on reaction, structure and properties of resulting geopolymer', *Ceramics International* Elsevier Ltd and Techna Group S.r.l., 37(2), pp. 533–541. doi: 10.1016/j.ceramint.2010.09.038.

Kumar, S., Kumar, R. and Mehrotra, S. P. (2010) 'Influence of granulated blast furnace slag on the reaction, structure and properties of fly ash based geopolymer', *Journal of Materials Science*, 45(3), pp. 607–615. doi: 10.1007/s10853-009-3934-5.

Lee, N. K. and Lee, H. K. (2015) 'Reactivity and reaction products of alkali-activated, fly ash / slag paste', *Construction and Building Materials*, 81, pp. 303–312.doi: 10.1016/j.conbuildmat.2015.02.022.

Marcin, M., Sisol, M. and Brezani, I. (2016) 'Effect of slag addition on mechanical properties of fly ash based geopolymers', *Procedia Engineering,* Elsevier B.V., 151, pp. 191–197. doi: 10.1016/j.proeng.2016.07.380.

Marjanovi, N. et al. (2014) 'Physical – mechanical and microstructural properties of alkali-activated fly ash – blast furnace slag blends', *Ceramics International*, 41(1), pp. 1421–1435. doi: 10.1016/j.ceramint.2014.09.075.

Marjanovi, N. et al. (2015) 'Comparison of two alkali-activated systems : mechanically activated fly ash and fly ash-blast furnace slag blends', 108, pp. 231–238. doi: 10.1016/j.proeng.2015.06.142.

Mucsi, G. (2016) 'Mechanical activation of power station fl y ash by grinding – A review', *Epitoanyag - Journal of Silicate Based and Composite Materials*, 68(2), pp. 56–61. doi: 10.14382/epitoanyag-jsbcm.2016.10.

Nath, S. K. et al. (2016) 'Microstructural and morphological evolution of fly ash based geopolymers', *Construction and Building Materials,* 111, pp. 758–765. doi: 10.1016/j.conbuildmat.2016.02.106.

Palomo, A. et al. (2007) 'Alkali-activated fly ash: Effect of thermal curing conditions on mechanical and microstructural development – Part II', *Fuel*, 86(3), pp. 315–322. doi: 10.1016/j.fuel.2006.07.010.

Puertas, F. et al. (2000) 'Alkali-activated fly ash/slag cements: Strength behaviour and hydration products' *Cement and Concrete Research*, 30(10), pp. 1625–1632. doi: 10.1016/S0008-8846(00)00298-2.

Shadnia, R. and Zhang, L. (2017) 'Experimental Study of Geopolymer Synthesized with Class F Fly Ash and Low-Calcium Slag', *Journal of Materials in Civil Engineering*, 29(10), p. 04017195. doi: 10.1061/(asce) mt.1943-5533.0002065.

Singh, B. et al. (2015) 'Geopolymer concrete : A review of some recent developments', *Construction and Building Materials*. Elsevier Ltd, 85, pp. 78–90. doi: 10.1016/j.conbuildmat.2015.03.036.

Toniolo, N. and Boccaccini, A. R. (2017) 'Fly ash-based geopolymers containing added silicate waste . A review', *Ceramics International*, Elsevier Ltd and Techna Group S.r.l., 43(17), pp. 14545–14551. doi: 10.1016/j.ceramint.2017.07.221.

Winnefeld, F. et al. (2010) 'Assessment of phase formation in alkali activated low and high calcium fly ashes in building materials', *Construction and Building Materials*, 24(6), pp. 1086–1093. doi: 10.1016/j. conbuildmat.2009.11.007.

Xu, H. and Van Deventer, J. S. J. (2000) 'The geopolymerisation of alumino-silicate minerals', *International Journal of Mineral Processing*, 59(3), pp. 247–266. doi: 10.1016/S0301-7516(99)00074-5.

Yao, X. et al. (2009) 'Geopolymerization process of alkali-metakaolinite characterized by isothermal calorimetry', *Thermochimica Acta*, 493(1–2), pp. 49–54. doi: 10.1016/j.tca.2009.04.002.

Yun-Ming, L. et al. (2016) 'Progress in materials science structure and properties of clay-based geopolymer cements: A review', *Progress in Materials Science*. Elsevier Ltd, 83, pp. 595–629. doi: 10.1016/j. pmatsci.2016.08.002.

9

Impacts of Municipal Solid Waste Heavy Metals on Soil Quality: A Case of Visakhapatnam

P.V.V. Prasada Rao and G. Siva Praveena
Andhra University, Visakhapatnam, India

CONTENTS

9.1 Introduction 69
9.2 Study Site History 70
9.3 Materials and Methods 70
    9.3.1 Sample Collection 70
    9.3.2 Metal Extraction Procedure 70
    9.3.3 Geo-Accumulation Index (I_{geo}) 70
9.4 Results and Discussion 71
    9.4.1 Lead 73
    9.4.2 Nickel 73
    9.4.3 I_{geo} 73
9.5 Conclusions 74
Acknowledgements 75
References 75

9.1 Introduction

Urbanization, expansion of economic activity and consumption patterns led to Municipal Solid Waste (MSW) generation (Schwarz-Herion et al., 2008) in huge quantities in urban areas. The relentless dumping of MSW is responsible for many insanitary conditions in many urban areas, challenging the comfortable existence of the human race. Landfills and open dumpsites are considered as some of the viable and only destination options for final disposal of waste generated globally (Phetyasone et al., 2018). It is evident that major metropolitans, including rural areas, are facing greater challenges in appropriate management and effective disposal of MSW generated due to awkward situations from source level to policy implementation, resulting in the failure of waste management. However, there are many factors that worsen the condition, which include unscientific practices adopted at the dumpsites without considering the impending effects (Schwarz-Herion et al., 2008). Climatic conditions of the region also play a crucial role in exacerbating the situation (Pickford,1977).

Conflicting ideas of authorities, and improper collection and transportation of waste from the source to the dumpsite affects the management activities leading to inconsistency at the other end (Ogundipe, 1978).

With the passage of time, the MSW produces a complex mixture of organic and inorganic fluid called leachate (Sampath Kumar & Swathi, 2014) which finds its way into the groundwater through soil sub-layers due to the pressure created by the continuous dumping activity. The leachate thus generated poses a severe threat to soil and ground water quality if not scientifically managed (Ikem et al., 2002). The heavy metals leached from the waste heaps often intensify the problem by contaminating both the groundwater and the soil strata affecting the contiguous sources. Many such studies recorded heavy metal concentrations exceeding the permissible levels (Aderemi et al., 2011; Magda & Gaber, 2015). These heavy metals are of both geogenic and anthropogenic origin, which are persistent leading to eco-toxicity

DOI: 10.1201/9781003217619-10

69

(Storelli et al., 2005). Earlier studies have indicated that Pb, Hg, Cd, Cu, Ni and Cr heavy metals are the major pollutants in soils near dumpsite areas (Esakku et al., 2003; Awokunmi et al., 2010). Modern day crisis of fertile soils provokes farmers and agriculturists to perceive dumpsite soils as one probable source of fertile soil enriched with high organic content. However, the use of these soils as an alternative to market-bought fertilizer is questionable due to the dynamicity of the waste dumped, enhanced metal burdens and practices adopted at the dumpsites (Partha et al., 2011).

9.2 Study Site History

The study area, Visakhapatnam, one of the fastest growing urban agglomerations of India, is nestled between Eastern Ghats and Bay of Bengal. The city is famous for its scenic beaches and is the deepest land-locked and protected natural port on the east coast of India. Greater Visakhapatnam is one of the major industrial corners serving the people in number of ways including migrants from adjoining Odisha state. The dumpsite, *"Kapuluppada"* is 25 kilometers away from Visakhapatnam city towards Vizianagaram and located between latitude 17°50′45 26″ N and longitude 83° 22′ 03 27′E in Kapuluppada village of Bheemunipatnam mandal. The dumpsite extends to about 100 acres, delimited by thick jade vegetation. The dumpsite receives daily 1,000–1,050 Metric ton/day of MSW generated in the city. The study area is dominated by foliated metamorphic rock called "Khondalites" (Varghese, 2012), which has its roots from the *"Khond"* tribe.

9.3 Materials and Methods

9.3.1 Sample Collection

The soil samples were collected randomly within the dumpsite. Samples were scooped from a depth of 15cms into air–tight polythene containers to sustain the nature of samples. The samples were shade-dried and ground into fine powder with the help of mortar and pestle and the contents passed through a 2mm sieve for metal analysis. Similar procedures were adopted for control soil samples, which were collected far away from the dumpsite location.

9.3.2 Metal Extraction Procedure

1gm of homogenized soil sample was digested with concentrated aqua-regia (1:3) for extraction of heavy metals as per standard procedures (NEPM, 2013; USEPA, 1996 3050B). During digestion, the contents were moistened with a little water and further heated. The soil sample in the acid mixture was digested on the hot plate until 5ml of the residue was left in the digestion flask, which was later allowed to cool down. The process was repeated until the digestion is complete and only 5ml residue remained in the digestion flask. The digested mixture was filtered through a Wattman filter paper, and the filtrate was made up to 50ml with de-ionized distilled water for further Atomic Absorption Spectroscopy (AAS) analysis.

9.3.3 Geo-Accumulation Index (I_{geo})

The Geo-Accumulation Index is used to assess the presence and the intensity of anthropogenic contaminant deposition on surface soil (Muller, 1979). A geo-accumulation index compares the measured concentration of the element (Cn) in the soil samples with the geochemical background value (Bn) and the constant 1.5 allows to analyze natural fluctuations in the content of a given substance in the environment and to detect the slightest anthropogenic influence.

$$\text{Geo-accumulation index} = \text{Log}_2 [\text{Cn}/1.5 \times \text{Bn}]$$

Based on the metal levels, the soil is classified into seven categories (Muller, 1981) ranging from Class 0 ($I_{geo} = 0$, unpolluted) to Class 6 ($I_{geo} > 5$, extremely polluted). The highest class (Class 6) reflects at least

Impacts of Municipal Solid Waste

TABLE 9.1

I_{geo} Classes

Value	Class	Soil Dust Quality(Muller, 1981)
$0 \leq$	0	Uncontaminated
$0 < I_{geo} < 1$	1	Uncontaminated to moderately contaminated
$1 < I_{geo} < 2$	2	Moderately contaminated
$2 < I_{geo} < 3$	3	Moderate to strongly contaminated
$3 < I_{geo} < 4$	4	Strongly contaminated
$4 < I_{geo} < 5$	5	Strong to very strongly contaminated
$I_{geo} \geq 5$	6	Very strongly contaminated

a 100-fold enrichment factor above background values. I_{geo} classification as indicated by Muller is shown in Table 9.1.

9.4 Results and Discussion

Soil samples collected from both dumpsite and control sites recorded a metal sequence of Pb > Ni. The concentration of lead ranges from 0.56mg/kg to 6.48mg/kg and 0.01mg/kg to 1.23mg/kg, while nickel ranges from 0.11mg/kg to 4.08mg/kg and 0.01mg/kg to 1.10mg/kg, in dumpsite and control soils respectively. The data is presented in Tables 9.2 and 9.3 and Figures 9.1 and 9.2.

TABLE 9.2

Heavy Metal Concentrations in Dumpsite Soil Samples

Dumpsite Sample No	Pb (mg/kg)	Ni (mg/kg)
S-1	4.22	0.11
S-2	1.20	1.26
S-3	2.12	0.66
S-4	0.56	1.48
S-5	1.26	3.20
S-6	4.11	3.24
S-7	4.80	2.64
S-8	2.42	2.16
S-9	1.82	1.11
S-10	3.44	0.84
S-11	3.17	0.64
S-12	2.14	0.33
S-13	1.85	3.40
S-14	3.33	2.46
S-15	4.16	2.15
S-16	4.26	0.85
S-17	6.48	1.23
S-18	2.16	1.11
S-19	2.44	2.20
S-20	3.82	4.08
Total	59.76	35.15
Average	2.98	1.75

TABLE 9.3

Heavy Metal Concentrations in Control Soil Samples

Control Sample No	Pb (mg/kg)	Ni (mg/kg)
S-1	BDL	BDL
S-2	0.01	BDL
S-3	0.20	0.04
S-4	1.00	0.01
S-5	BDL	BDL
S-6	1.23	0.20
S-7	0.10	BDL
S-8	0.40	0.20
S-9	0.10	0.20
S-10	BDL	0.12
S-11	BDL	BDL
S-12	BDL	0.13
S-13	0.10	BDL
S-14	1.20	1.10
S-15	0.20	0.03
S-16	BDL	BDL
S-17	0.30	BDL
S-18	0.10	0.04
S-19	BDL	0.02
S-20	0.12	0.11
Total	5.06	2.20
Average	0.25	0.11

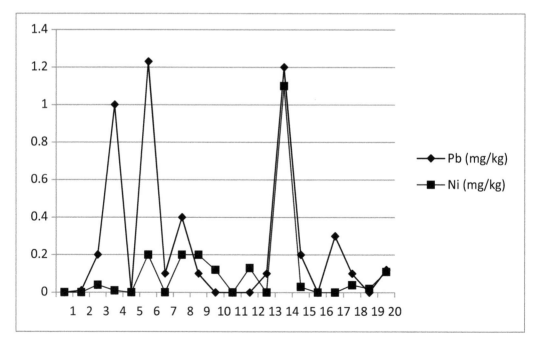

FIGURE 9.1 Metal concentrations of control samples (mg/kg).

Impacts of Municipal Solid Waste

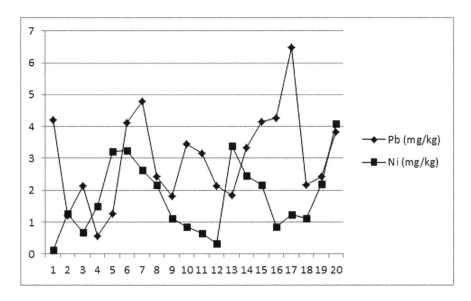

FIGURE 9.2 Dumpsite soil metal concentrations (mg/kg).

9.4.1 Lead

Lead levels and its contamination are some of the major soil issues in recent years owing to its toxic nature affecting plant, animal and human life. The element finds its way into the soil strata through the paints, batteries, smelting operations, e-waste and vehicles using leaded petrol. The mean Pb level (2.98mg/kg) presented in Table 9.2 of dumpsite soil is slightly above the permissible levels 2mg/kg of the World Health Organization (WHO) (Hasan et al., 2012). The control soil sample recorded metal levels well within the recommended levels by WHO. However, the incessant dumping activity might worsen the soil condition beyond revamp. Contamination of groundwater sources has also been reported in and around the study site which point towards extent of leachate percolation from the waste heaps (Sampath Kumar & Swathi, 2014). In addition, Pb is considered a human carcinogen causing neurological disorders (Fatoki et al., 2005), which calls for an immediate attention to mitigate. Also, the thoughtless practices like open burning practiced at the dumpsite amplify the situation even more where the heavy metal particulates enter the air leading to air pollution.

9.4.2 Nickel

Ni is essentially at minimal levels, yet the pollution caused by it annoys since it is involved in the growing decline of microorganisms that play a vital role in soil fertility and a major bio-indicator species of soil quality. The disposal of waste electronic goods into the MSW disposal site leads to elevated Ni levels in the dumpsite soils. In the present study, the mean Ni level (1.75mg/kg) presented in Table 9.2 was within the permissible levels (10mg/kg) by WHO, while the regular burning activity of e-waste (Figure 9.4) for metal is one way of exposure to Ni (Lenntech, 2009). Nickel, generally, is adsorbed by the sediments and the soil particles thereby become immobile. However, the acidic conditions in the dumpsite would result in metal leaching contaminating the soil.

9.4.3 I_{geo}

The geo-accumulation index calculated for the mean metal levels of soil samples against the shale values indicated that the soil samples of the study area are very strongly contaminated with both Pb and Ni falling into class 6 of the soil dust quality as shown in Table 9.1 and Figure 9.3. Further, the level of biogenic

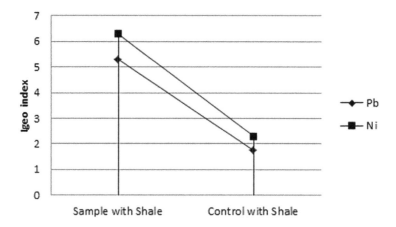

FIGURE 9.3 I_{geo} index of dumpsite and control samples against shale values.

FIGURE 9.4 View of smoldering activity at the dumpsite.

and anthropogenic contributions of Pb and Ni in the dumpsite soil samples were compared with a geologically similar material (Tijan et al., 2004) collected from a pristine locality away from the dumpsite. The control samples against shale values also reported considerable metal levels falling into class 2 and 3 for Pb and Ni presented in Table 9.1 and Figure 9.3 which may be due to the natural geological weathering process. However, mean metal levels were well within the permissible standards of WHO recommended for plants.

9.5 Conclusions

The I_{geo} index calculated revealed that concentrations of Pb and Ni in the dumpsite soil samples, are above the permissible levels and the contamination is at varying degrees. This is an impending threat to the other vital resources like ground water and vegetation. Lack of proper e- waste disposal in the study area could exacerbate the soil quality further with the dynamicity of the leachate generated. Although the mean metal levels recorded in the present study falls under moderately contaminated range and are not really beyond the permissible levels for soil, the metal concentrations recorded high for few samples, which specifies for a continuous monitoring and assessment in the study area to identify the altering chemistry of the dumpsite soils and to take appropriate remedial measures. A recent proposal of *"Bio-mining"* in the present dumpsite for metal leaching from the waste is temperature dependent and would entail temperature resistant microbial species. However, this activity would not survive microbes with the continual smoldering activity at the dumpsite eventually affecting the soil composition. The obtained "secondary soil" would often affect crop growth which still requires an additional testing prior to offsite application.

Impacts of Municipal Solid Waste

Thus, containment of soil pollution needs comprehensive strategies including strict law enforcement within and around the dumpsites.

ACKNOWLEDGEMENTS

One of the authors, Ms. G. Siva Praveena, is grateful to UGC, New Delhi for providing financial assistance in the form of Rajiv Gandhi National Fellowship. Thanks are due to the Department of Environmental Sciences, Andhra University for providing required infrastructural facilities to carry out the present work.

REFERENCES

Aderemi, A.O., Oriaku, A.V., Adewumi, G.A., & Otitoloju, A.A. (2011). Assessment of groundwater contamination by leachate near a municipal solid waste landfill, *African Journal of Environmental Science and Technology*, 5, pp. 933–940.

Awokunmi, E.E., Asaolu, S.S., & Ipinmoroti, K.O. (2010). Effect of leaching on heavy metals concentration of soil in some dumpsites. *African Journal of Environmental Science and Technology*, 4(8) pp. 495–499.

Esakku, S., Palanivelu, K., & Joseph, K. (2003). Assessment of heavy metals in a municipal solid waste dumpsite. Workshop on Sustainable Landfill Management, 3–5 December, Chennai, India, pp. 139–145.

Fatoki, O.S., Lujiza, N., & Ogunfowokun, O.A. (2005). Trace metal pollution in Umtata river, *Water S.A.*, 28, (2), 183.

Hasan, Z., Anwar, Z., Khattak, K.U., Islam, M., Khan, R.U. and Khattak, J.Z.K. (2012). Civic pollution, and its effect on water quality of River Toi at District Kohat, NWFP, *Research Journal of Environmental and Earth Sciences*, 4, pp. 5.

Ikem, A.O., Osibanjo, M.K., Sridhar, C., & Sobande, A. (2002). Evaluation of groundwater quality characteristic near two waste sites in Ibadan and Lagos, Nigeria, *Water, Air, and Soil Pollution*, 140: pp. 307–333.

Lenntech, W.T. (2009). Chemical properties, health, and environmental effects of copper, Lenntech Water Treatment and Purification Holding B.V.

Magda, M.A., & Gaber, I.A. (2015). Impact of landfill leachate on the groundwater quality: A case study in Egypt, *Journal of Advanced Research* 6, pp. 579–586.

Muller, G. (1979). Schwermetalle in den sedimenten des Rheinse Veranderungenseitt 1971. Umschau 1979: pp. 778–783.

Muller, G. (1981). DieSchwermetallbelastung der sedimente des Neckars und seiner Nebenflusse: eineBestandsaufnahme, *Chem Ztg*, 105, pp. 157–164.

NEPM. (2013). Amendment of the assessment of site contamination, 2013.

Ogundipe, S. (1978). Problems of SOLID WASTES in Ibadan: A rejoinder, *Daily Sketch (Ibadan)*, 1978, pp. 5 and 11.

Partha, V., Murthya, N.N., & Saxena, P.R. (2011). Assessment of heavy metal contamination in soil around hazardous waste disposal sites in Hyderabad city (India): natural and anthropogenic implications, *Journal of Environmental research and management*, 2 (2), pp. 027–034.

Phetyasone, X., Jiro, T., Chart C., and Tanchuling, M.A.N. (2018). Characterization of landfill leachates and sediments in major cities of Indochina peninsular countries—heavy metal partitioning in municipal solid waste leachate, *Environments*, 5, 65.

Pickford, J. (1977). Solid waste in hot climate, In Feachhem, R., McGary, M. and Mara, D. (eds.) *Water and Health in Hot Climate*, 1977, John Wiley & Sons, London.

Sampath Kumar, M.R.S., & Swathi G., (2014). Unplanned municipal solid waste dumps and their impact on water quality - a case study from Visakhapatnam, Andhra Pradesh, South India, *IOSR Journal of Environmental Science, Toxicology and Food Technology*, 8 (9), pp. 1–5.

Schwarz-Herion, O., Omran, A., & PRapp, H. (2008). A case study on successful municipal solid waste management in industrialized countries by the example of Karlsruhe City, Germany, *Journal of Engineering Annals, of the Faculty of Engineering Hunedoara*, 6 (3) pp. 266–273.

Storelli, M., Storelli, A., D'Addabbo, R., Marano, C., Bruno, R., & Marcotrigiano, G. (2005). Trace elements in loggerhead turtles (Caretta caretta) from the eastern Mediterranean Sea: overview and evaluation, *Environmental Pollution*, 135(1), 163–170.

Tijani, M. N., Jinno, K., & Hiroshiro, Y. (2004). Environmental impact of heavy metal distribution in water and sediment of Ogunpa River, Ibadan area, southwestern Nigeria. *Journal of Mining and Geology*, 40(1), 73–83.

USEPA. (1996). Acid digestion of sludges, solids and soils, USEPA 3050B, In SW-846 Pt 1; Office of Solid and Hazardous Wastes, USEPA: Cincinnati, OH.

Varghese, P.C. (2012). *Engineering Geology for Civil Engineers*. PHI Learning Pvt. Ltd. pp. 126.

10

Effective Utilization of Industry Solid Waste into the Concrete and Its Management

V.S. Vairagade
Priyadarshini College of Engineering, Nagpur, India

B.V. Bahoria
Priyadarshini College of Engineering, Nagpur, India

Rakesh Patel
Priyadarshini College of Engineering, Nagpur, India

P.T. Dhorabe and V.R. Agrawal
Priyadarshini College of Engineering, Nagpur, India

N.P. Mungle
Priyadarshini College of Engineering, Nagpur, India

CONTENTS

10.1 Introduction 77
10.2 Materials Properties 78
    10.2.1 Cement 78
    10.2.2 Aggregates 78
    10.2.3 Sugar Cane Bagasse Ash 78
    10.2.4 Marble Slurry Dust 79
10.3 Mixture Proportioning 79
10.4 Experimental Methodology 80
    10.4.1 Test on Fresh Concrete 80
    10.4.2 Test on Hardened Concrete and Mortar 80
    10.4.3 Compressive Strength of Concrete 81
10.5 Experimental Results and Discussions 81
    10.5.1 Workability of Fresh Concrete 81
    10.5.2 Compressive Strength of Concrete 82
    10.5.3 Compressive Strength of Mortar 83
10.6 Conclusions 85
References 85

10.1 Introduction

Cement concrete is the most extensively used construction material in the world. Ordinary Portland Cement (OPC) is recognized as the major construction material throughout the world. Portland cement is responsible for about 5–8% of global CO_2 emission. This environmental problem will most likely be increased due to exponential demand of Portland cement (Aitcin, 2000). Industrial wastes, such as rice husk ash, fly ash and silica fume are being used as supplementary cement replacement materials (Sathawane et al. 2013). In addition to these, agricultural wastes such as rice husk ash, wheat straw

DOI: 10.1201/9781003217619-11

ash, and sugarcane bagasse ash are also being used as pozzolanic materials, and hazel nutshell used as cement replacement material (Somna et al., 2012). India is the second largest producer of sugarcane and large quantity of bagasse ash (67,000 tonnes/day) and a large quantity of sugarcane bagasse is available from sugar mills (Ganesan et al., 2007). Sugarcane bagasse ash is a byproduct of sugar factories, and it is produced by burning sugarcane bagasse. It was found that SCBA improves the properties of concrete and mortar such as compressive strength, water tightness in some percentage of replacement (Vairagade & Sathawane, 2013). Initiatives are taken worldwide to control and to manage the agricultural waste by replacing it with cement to make a green environment (Montakarntiwong et al., 2013). There are various studies related to use of SCBA as supplementary cementitious material in concrete and mortar (Bahurudeen et al., 2015), (Nuntachai & Chai, 2009), (Modania & Vyawahare, 2013).

One of the major wastes produced in the stone industry during cutting, shaping, and polishing of marbles is marble dust. Past studies related to use of marble slurry dust as supplementary cementitious material in concrete and mortar shows that it is promising material (Topcu et al., 2009), (Karasahin, & Terzi, 2007). Due to the availability of the large quantity of waste produced, this paper deals with the possible use of sugarcane bagasse ash and marble slurry dust in concrete and mortar as partial replacement of cement.

10.2 Materials Properties

For this research work, cement, sand, coarse aggregate, water, sugarcane bagasse ash and marble slurry dust were used.

10.2.1 Cement

The cement used was OPC (43 Grade) with a specific gravity of 3.15. Initial and final setting time of the cement was 20 mins and 227 mins, respectively. OPC 43 grade used in this experimentation conforming to IS-8112, 1989.

10.2.2 Aggregates

Good quality river sand was used as a fine aggregate. The specific gravity was 2.45. Coarse aggregate passing through 20mm and retained 12mm sieve was used. Its specific gravity and dry density was 2.67 conforming to IS-383, 1970.

10.2.3 Sugar Cane Bagasse Ash

Sugarcane bagasse ash used in this research was obtained from Purti Power Plant, Bela, Nagpur, Maharashtra, India. Sugarcane bagasse ash is a byproduct of sugar factories, and it is produced by burning sugarcane bagasse. The bagasse ash passes from a 90µm sieve and retained on 45µm; the retained bagasse ash is taken for the preparation of concrete and mortar. Figure 10.1 shows the raw sugarcane bagasse and sugarcane bagasse ash.

FIGURE 10.1   (a) Raw sugarcane bagasse, (b) Sugarcane bagasse ash.

Utilization of Industry Solid Waste

FIGURE 10.2 Marble slurry dust.

TABLE 10.1

Chemical Composition of Sugar Cane Bagasse Ash and Marble Slurry Dust

Sr. No	Oxides	Mass (gm/100gm) SCBA	Mass (gm/100gm) MSD
1	Silicon Oxide (SiO_2O_2)	53.44	5.97
2	Aluminum Oxide (Al_2O_3)	14.73	0.35
3	Ferrous Oxide (Fe_2O_3)	11.41	2.87
4	Calcium Oxide (CaO)	3.45	36.48
5	Magnesium Oxide (MgO)	6.77	12.02
6	LOI	10.30	38.06

10.2.4 Marble Slurry Dust

Figure 10.2 shows marble powder collected from the dressing and processing unit in Nagpur, Maharashtra, India. The marble dust used in this project is in powdered form, odorless, grey in color with the moisture content of 1.59%. Marble is a metamorphic rock resulting from the transformation of a pure limestone. The purity of the marble is responsible for its color and appearance: it is white if the limestone is composed solely of calcite (100% $CaCO_3$).

In order to be used as a mineral admixture in mortar and concrete, material must have appropriate chemical properties. The moisture content of bagasse and marble slurry, weight and the amount of ash were measured first. The chemical compositions of bagasse ash and marble slurry dust were investigated and compared with OPC based on limitation given by IS 1489 (Part-I), 1991.

The results of chemical composition of sugar cane bagasse ash and marble slurry dust are shown in Table 10.1.

It was found that silicon dioxide is 54% in sugar cane bagasse ash as a main oxide. The other oxides present are aluminum oxide 14(g/100g), iron oxides of about 11(g/100g) and CaO of about 3(g/100g). MnO was found to be about 6.7(g/100g) and loss on ignition (LOI) was found to be 10.30(g/100g) composition. The similar oxide has been found with that of cement.

In the case of marble slurry dust, it was found that calcium oxide content is major oxide which is nearly 37% by weight of marble dust powder. Also, the lime content in marble dust is more than any other component.

10.3 Mixture Proportioning

The mixture proportioning was done according to the Indian Standard Recommended Method IS 10262, 2009 and with reference to IS 456, 2000. The total binder content was 438.00kg/m³, fine aggregate was taken as 673.29kg/m³ and coarse aggregate as 1073.35kg/m³. The water to binder ratio was kept constant

TABLE 10.2

Concrete Mix Proportions

Material	Quantity (kg/m³)	Proportion
Cement	438 kg/m³	1
Sand	673.29 kg/m³	1.537
Coarse Aggregates	1073.35 kg/m³	2.45
Water	197 kg/m³	0.45

as 0.45. The total mixing time was 5 minutes, the samples were then casted and left for 24 hours before demolding. They were then placed in the curing tank until the day of testing. Cement, sand and coarse aggregate were properly mixed together in the ratio 1:1.537:2.45 by weight before water was added and properly mixed together to achieve homogenous material. Water absorption capacity and moisture content were taken into consideration. Cube molds were used for casting. Compaction of concrete in three layers with 25 strokes of 16mm rod was carried out for each layer. The concrete was left in the mold and allowed to set for 24 hours before the cubes were demolded and placed in curing tank. The concrete cubes were cured in the tank for 7, 21 and 28 days.

Mix proportion for concrete for tested material is shown in Table 10.2.

10.4 Experimental Methodology

10.4.1 Test on Fresh Concrete

Concrete is said to be workable when it is easily placed and compacted homogeneously. Workability is one of the physical parameters which affects strength and durability as well as cost of labor and appearance. Fresh concrete was tested using slump cone test to find the workability of control concrete and concrete with sugarcane bagasse ash and marble slurry dust at 0%, 5%, 10%, 15% and 20% replaced with ordinary cement.

The effect of workability in terms of slump values is shown in Figure 10.3.

10.4.2 Test on Hardened Concrete and Mortar

The experiments were carried out in two sets. In first set of experiments, compressive strength of concrete with different blends at 0%, 5%, 10%, 15% and 20% partial replacement by sugarcane bagasse ash and marble slurry dust were carried out. The water to binder ratio was kept constant to 0.45 for all the blends.

FIGURE 10.3 Slump cone test.

Utilization of Industry Solid Waste 81

In a second set of experiments, compressive strength of mortar with different blends at 0%, 5%, 10%, 15% and 20% partial replacement by sugarcane bagasse ash and marble slurry dust were carried out. The water to binder ratio was kept constant to 0.4 and aggregate to binder ratio fixed to 1:3 for all the blends.

10.4.3 Compressive Strength of Concrete

The strength of concrete is usually defined and determined by the crushing strength of 150mm × 150mm × 150mm, at ages of 7, 21 and 28 days. It is the most common test conducted on hardened concrete as it is an easy test to perform and also most of the desirable characteristic properties of concrete are qualitatively related to its compressive strength. Steel mold made of cast iron dimension 150mm × 150mm × 150mm was used for casting of concrete cubes filled with concrete as 0%, 5%, 10%, 15% and 20% sugarcane bagasse ash and marble slurry dust replace with ordinary cement. The mold and its base rigidly damped together so as to reduce leakages during casting. The sides of the mold and base plates were oiled before casting to prevent bonding between the mold and concrete. The cube was then stored for 24 hours undisturbed at temperature of 18°C to 22°C and a relative humidity of not less than 90%.

It also stated in IS-516, 1959 that the load was applied without shock and increased continuously at the rate of approximately 140kg/cm^2/min until the resistance of specimen to the increasing loads breaks down and no greater load can be sustained. The maximum load applied to the specimen was then recorded as per IS-516, 1959.

In each category, three cubes were tested, and their average value reported.

The compressive strength was calculated as follows:

$$\text{Compressive strength (MPa)} = \text{Failure load/cross sectional area.}$$

10.5 Experimental Results and Discussions

10.5.1 Workability of Fresh Concrete

The results of workability in terms of slump values are shown in Figure 10.4.

Figure 10.4 shows the test results for effects of replacement of sugarcane bagasse ash and marble slurry dust on workability of concrete. The slump for conventional concrete was found to be 90mm, whereas

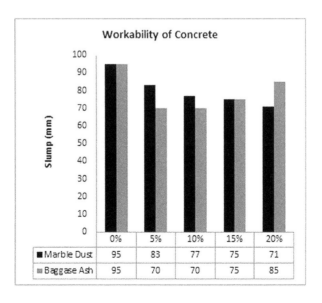

FIGURE 10.4 Effect of sugarcane bagasse ash and marble slurry dust on workability of concrete.

addition of bagasse ash to 5% and 10% has a slump of 70mm and for 15% replacement it was 75 mm. The analysis shows the addition of SCBA to 5% and 10% decreases slump, whereas at 15% it was found to have increased to 75mm and the further replacement of Sugarcane Bagasse Ash to 20% resulted in an increase in the slump to 85mm. It shows that increase in SCBA in concrete increases the workability.

In the case of marble dust in concrete, the workability decreases as replacement increases. For 5% replacement of marble dust in concrete does not affect the workability so much. But after further replacement, workability decreases significantly.

10.5.2 Compressive Strength of Concrete

The results of the compressive strength tests for blends containing 0%, 5%, 10%, 15% and 20% sugarcane bagasse ash and marble slurry dust replace with ordinary cement at various ages are shown in Figure 10.5, 10.6 and 10.7 below.

It was observed that, optimum levels for replacement of SCBA in concrete was found to be 15% and 10% for MSD. Further replacement of cement by SCBA and MSD shows decrease in strength of concrete. The concrete with SCBA has 23% higher compressive strength for 15% replacement of SCBA

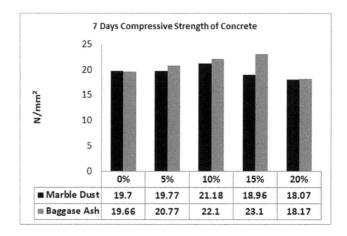

FIGURE 10.5 7 Days compressive strength of concrete.

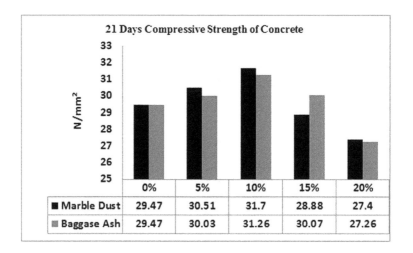

FIGURE 10.6 21 Days compressive strength of concrete.

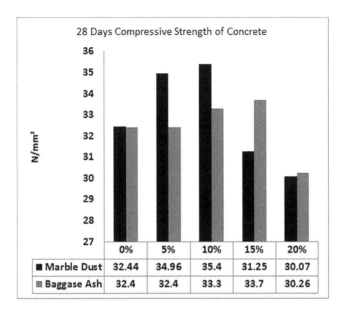

FIGURE 10.7 28 Days compressive strength of concrete.

for 7 days than conventional concrete whereas the concrete with SCBA has 12% higher compressive strength for 15% replacement of SCBA for 21 days than conventional concrete and 4% higher compressive strength for 15% replacement for 28 days than conventional concrete.

In case of replacement with marble slurry dust, the 7 days compressive strength of concrete by 5% and 10% replacement was increased by 1% and 1.07%, respectively. But after 15% and 20% replacement the strength decreased by 0.98% and 0.92%, respectively. 21 days compressive strength of concrete by 5% and 10% replacement was increased by 1.03% and 1.07% respectively. But, after 15% and 20% replacement, the strength decreased by 0.98% and 0.93% respectively. 28 days compressive strength of concrete by 5% and 10% replacement was increased by 1.07% and 1.09% while, after 15% and 20% replacement compressive strength of concrete decreased by 0.96% and 0.92% respectively. So, after 15% and 20% replacement the graph moves downward, meaning the compressive strength decreases after 15% and 20% replacement of marble dust powder in concrete. The compressive strength for 5% and 10% replacement increases and at 10% replacement of marble dust in concrete by cement gives maximum strength. So, the replacement of marble dust powder in cement should be done up to 10%.

10.5.3 Compressive Strength of Mortar

Compressive strength of mortar results for blends containing 0%, 5%, 10%, 15% and 20% sugarcane bagasse ash and marble slurry dust replace with ordinary cement at various ages are shown in Figure 10.8, 10.9 and 10.10 below.

It was observed that, optimum level of replacement was found to be 10% of for both materials. Further replacement of cement by SCBA and MSD shows decrease in strength of concrete.

It was observed that maximum compressive strength for mortar is found to be 25.25N/mm^2 for 28 days replacement, which decreases at 28 days of curing. The compressive strength of 7 and 21 days was equal for 20% replacement of cement with SCBA. The strength of mortar was found to significantly increase at 21 days of curing.

The mortar with SCBA has 22% higher compressive strength for 15% replacement of SCBA for 7 days of curing, 20% higher compressive strength for 15% replacement of SCBA for 21 days of curing, and 9.44% higher compressive strength for 20% replacement for 28 of curing days than conventional mortar.

84   *Circular Economy in the Construction Industry*

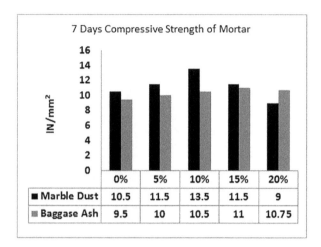

FIGURE 10.8 7 Days compressive strength of mortar.

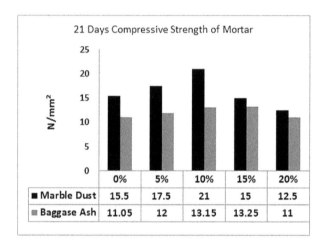

FIGURE 10.9 21 Days compressive strength of mortar.

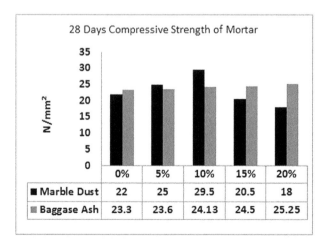

FIGURE 10.10 28 Days compressive strength of mortar.

Utilization of Industry Solid Waste   85

10.6 Conclusions

This paper focuses on use of SCBA as a pozzolanic material to produce green concrete. Both waste materials can be used to in concrete and mortar as there is always a need to overcome the problem of disposal of industrial waste and greenhouse effect.

On the basis of experimental investigation, the following conclusions were drawn.

1. SCBA can be a good replacement for cement in concrete as well as mortar compared to MSD. Therefore, SCBA is a more promising material than MSD.
2. The SCBA concrete gives higher compressive strength compared to MSD than that of control concrete.
3. Replacement of cement by SCBA increases workability of fresh concrete whereas MSD reduces the workability of concrete.
4. The compressive strength was found to be 17.45% more at 15% replacement for 7 days of curing, 6.06% more at 10% replacement for 21 days of curing and 4.01% more at 15% replacement for 28 days of curing than conventional concrete.
5. Bagasse ash can prove to be a potential ingredient of concrete since it can be an effective replacement to cement.

REFERENCES

Aitcin, P.C., "Cements of yesterday and today: Concrete of tomorrow", *Cement and Concrete Research*, Vol. 30, 2000, pp. 1349–1359.

Bahurudeen A., Kanraj D, Gokul V, Santhanam M, "Performance evaluation of sugarcane bagasse ash blended cement in concrete", *Cement & Concrete Composites*, Vol. 59 (2015), pp. 77–88.

Ganesan K, Rajagopal K, Thangavel K.," Evaluation of bagasse ash as Supplementary cementitious material", *Cement and Concrete Composites*, 2007; 29: 515–524.

IS 456-2000, Code of practice for plain and reinforced concrete, Bureau of Indian Standards, New Delhi, India

IS 10262-2009, Recommended guidelines for concrete mix design, Bureau of Indian Standards, New Delhi, India

IS 383-1970, Specification for coarse and fine aggregates from natural sources for concrete (second revision), Bureau of Indian standards, New Delhi, India.

IS 8112-1989, 43 grade ordinary portland cement – specification (first revision), IS 8112:1989, Bureau of Indian Standards, New Delhi.

IS 1489 (Part-I)-1991, Portland - Pozzolana cement specification, Bureau of Indian Standards, New Delhi, India

IS: 516-1959, Indian standard methods of tests for strength of concrete, Bureau of Indian Standards, New Delhi, India

Karasahin, M., and Terzi, S., "Evaluation of marble waste dust in the mixture of asphaltic concrete". *Construction and Building Materials*, Vol. 21, 2007, pp. 616–620.

Modania P., Vyawahare M., "Utilization of bagasse ash as a partial replacement of fine aggregate in concrete", *Proceedia Engineering*, Vol. 51, 2013, pp. 25–29.

Montakarntiwong K, Chusilp N, and Tangchirapat W., "Strength and heat evolution of concretes containing bagasse ash from thermal power plants in sugar industry", *Materials and Design*, Vol. 49, 2013, pp. 414–420.

Nuntachai C, Chai J., "Utilization of bagasse ash as a pozzolanic material in concrete"s, *Construction and Building Materials*, Vol. 23, 2009, pp. 3352–3358.

Sathawane S., Vairagade V., and Kene K., "Combine Effect of Rice Husk Ash and Fly Ash on Concrete by 30% Cement Replacement", Published by Elsevier Ltd., Science Directs, Procedia Engineering, Vol. 51, 2013, pp. 35–44.

Somna R, Jaturapitakkul C, Rattanachu P, Chalee W., "Effect of ground bagasse ash on mechanical and durability properties of recycled aggregate concrete", *Materials and Design* 2012; 36: 597–603.

Topcu I, Bilir T, Tayfun Uygunog~lu, "Effect of Waste Marble Dust Content as Filler on Properties of Self-Compacting Concrete". *Construction and Building Materials*, Vol. 23, 2009, pp. 1947–1953.

Vairagade V. and Sathawane S., "Investigation on behavior of concrete by partial replacement of cement with rice husk ash and fly ash using steel fiber", *Proceedings of International Conference on Innovations in Concrete Construction organized by UKIERI Concrete Congress* held at Jalandhar, Punjab, 5–8 March 2013, pp. 1462–1473.

11

Utilization of Industrial Waste in Normal Concrete: A Review

Srishti Saha, Tribikram Mohanty, and Bitanjaya Das
KIIT Deemed to be University, Bhubaneswar, India

CONTENTS

11.1 Introduction 87
11.2 Fresh Property of Waste Materials 88
     11.2.1 Workability of Industrial Waste 88
11.3 Mechanical Property 89
     11.3.1 Effect of Waste Materials on Compressive Strength of Concrete 89
     11.3.2 Effect of Waste Materials on Tensile Strength and Flexural Strength of Concrete 91
11.4 Conclusion 93
References 94

11.1 Introduction

Concrete is one of the most used materials for construction globally. In India, there is a progressive increase in the investment in infrastructure development which generated a huge demand for construction materials across the country. Currently, the field of concrete technology has witnessed many efforts of exploration and research on the use of by-products of the industrial sector and waste resources for concrete manufacture. Proper utilization of waste materials can decrease the cost of concrete production, improve the characteristics of fresh and hardened concrete, and decrease the environmental impact [1, 2]. A major component of concrete is contributed by the cement; hence it will be more efficient, effective, durable and energy-saving if industrial wastes are utilized as the cement substitution in concrete. Cement manufacturing industries contribute 5% of the total carbon dioxide (CO_2) emission, in which 50% is emitted due to chemical actions and 40% due to fuel burning. It has been studied that 1 tonne of cement produces 1 tonne of carbon dioxide. As the demand for cement will keep increasing with time, it is essential to use alternative cementitious materials in order to reduce carbon emission. It was observed that concrete manufactured using waste materials and by-products of the industry have superior qualities as compared to conventionally made concrete. Some harmful atmospheric gases, like CO_2, methane, chlorofluorocarbon, and nitrous oxide, absorb the outward infrared radiation, hence warming the atmosphere. So, as the concentrations of these harmful gases are increasing daily, so our ecosystem is also engrossing additional sunlight resulting in the increase of global temperature [3–5]. The amounts of carbon dioxide emission in the cement industry are 900kg for every 1,000kg of cement produced. In order to decrease the greenhouse effect, decrement in the cement quantity produced in the construction products, as cement is used as a major source in concrete which is harmful and a cause of the greenhouse effect. The demand for concrete will continue to increase daily which makes the use of industrial by-products and other alternative material in concrete manufacture urgent. Pozzolanic materials which are mixed in blended cement are mainly aluminous or siliceous or non-siliceous material. These materials themselves have no cementitious nature in the normal state but upon reaction with calcium hydroxide under moist conditions at normal temperature lead to the development of cementitious properties [6].

DOI: 10.1201/9781003217619-12

FIGURE 11.1 (a) Silica fume (b) Ferrochrome ash (c) Red mud (d) GGBS (e) Ferrochrome slag (f) Marble dust (g) Recycled brick (h) Mortar (i) Waste ceramic (j) Rice husk ash.

TABLE 11.1
Physical Properties of Different Types of Materials [7–12]

Physical Properties	OPC	Ferrochrome Ash (FA)	Silica Fume (SF)	Red Mud (RM)	GGBS	Marble Dust
Specific gravity	3.15	2.24	2.22	2.99	2.90	2.67
Fineness (Blaine's permeability method) (m²/kg)	370	571	30,000	20,000	463	350
Color	Dark gray	Gray	Light to dark gray	Brown	Off-white	White
PH	11	9.79	12.63	10.43	9.2	11.32
Density (g/cm³)	3.10	2.24	0.6	2.70	2.88	2.67

TABLE 11.2
Physical Properties of Different Types of Recycling Materials [13–16]

Physical Property Recycling Materials	Specific Gravity	Water Absorption (%)	Bulk Density (kg/m³)	Porosity (%)	Fineness (Blaine's Permeability Method) (m²/kg)
Mortar	2.01	3.9	1,570–1,650	-----	0.00051
Recycled Brick	1.702	15.81	1,805	48.6	0.00025
Waste ceramic	2.300	5.5	2,263	-----	0.00079
Precast	2.42	33.71	2,640	21	0.00042

Hence, the objective of this study is to investigate the performance of cement concrete as supplemented with industrial waste. The aim of this paper is to utilize the industrial wastes in normal concrete. Different types of industrial waste are shown in Figure 11.1 (a-j). and Tables 11.1, 11.2 and 11.3 show the physical properties of different types of waste.

11.2 Fresh Property of Waste Materials

11.2.1 Workability of Industrial Waste

Workability is measured, in general, by slump testing. It can be described as a consistency measure. Workability of Ground Granulated Blast furnace Slag (GGBS), silica fume, recycled brick, ceramic, fly ash, glass waste, recycled aggregate (RA), etc. depends on the curing age, sensitivity to curing, the heat of

Utilization of Industrial Waste

TABLE 11.3

Physical Properties of Ferrochrome Slag [3, 9, 16]

Physical Property	Patro et.al	Panda et.al (2013)	Das et.al (2014)	Maximum Limit (%)	References
Specific gravity	2.79	2.84	3.21	5.00	
Water absorption (%)	0.80	0.42	0.80	2.00	MORTH (5th Rev. P 267)
Flakiness index (%)	14.95	-----	9.28	35	MORTH (5th Rev. P 442)
Elongation index (%)	10.50	------	14.45		MORTH (5th Rev. P 267)
NB: MORTH = Ministry of Road Transport & Highways					

hydration, porosity, shape, and size of the aggregate [7–12]. It can be described as a consistency measure. Workability is the property of fresh concrete. Depending upon the type of crusher used to shape and size of the aggregate this also affect workability. As compared to round aggregate, annular shaped aggregate has higher contact area and this behavior also enables use of the reclaimed waste as precast concrete, recycled brick, mortar, and recycling plant or rubble. As the contact area is high, the absorption of water is also high, therefore, the water-cement ratio will become higher in order to obtain similar workability. Workability of ceramic waste, waste clay brick, waste mortar depends on the water absorption, porosity, and pore size of aggregate is the main reason behind the reduction of slump value of waste aggregate comparison to normal concrete. It was observed that when coarse and fine recycled aggregate mixed with ceramic brick workability is improved by a significant value. For 100% replacement slump the value was increased up to 4.6% and 7% for normal concrete and ceramic brick, respectively [13, 14]. The slump values of the waste materials slightly increase with an increase in saline content. It was suggested that replacement of natural aggregate (NA) by recycled aggregate (RA) requires similar characteristics, which can be attained by using a different process like crushing equipment. For the achievement of the same compressive strength and workability as conventional concrete, it requires to add more cement to concrete [15–17].

It was observed that angular and elongated shape of aggregates shows higher slump value, as well as the water/cement ratio was an efficient property to achieve the workability. Marble waste as a basalt and granite sand concrete, river sand has directly affected by this property [11, 12].

It was observed that at 20% replacement of GGBS strength increased at curing age of 28 days. Up to 40% addition of GGBS workability will be normal. It was found that for cement, GGBS can be used as a substitute material which can minimize consumption of cement and the construction cost [5, 10]. When different pozzolanic materials like GGBS, fly ash, etc. are partially replaced to the silica fume, workability of silica fume was better at curing age 28 days than that of other materials [1, 4, 5, 9]. The proportions of silica fume used ranged up to 20%. It was also certified that with the adding of silica fume and marble waste, workability of concrete should be reduced. Using water-reducing agent as a superplasticizer, workability can be solved from an economic point of view [2, 10].

11.3 Mechanical Property

11.3.1 Effect of Waste Materials on Compressive Strength of Concrete

Strength durability and structural performance of concrete are affected by the compressive strength of concrete, therefore, becoming one of the most effective properties of hardening of concrete. Properties of industrial waste depend upon various factors, for example, water/cement ratio, use of super-plasticizer, mineral dust content, water absorption, curing condition [1–5]. It was observed that cement had been replaced partially with fly ash. The proportions of replaced fly ash varied from 0–50%. The early development of strength, that is, for first 3– to 7 days in general, witnesses a decrement upon the addition of greater proportions of fly ash, which could be due to free lime present in concrete that keeps on reacting during the early curing phase. At 56 days, the ultimate strength of concrete is attained. The increasing strength for the cement mortar comprising coarse fly ash required a larger duration of time in comparison to the reference sample as cement mortars increased with the decrement of the particle size of fly ash [2]. It was found that up to 20% replacement of GGBS, red mud strength increases at age of 28 days as concrete is more compact at the high finer material with a smaller quantity of voids gives higher strength. Silica fume shows higher strength for longer concrete ages up to 25% replacement of cement [5, 8, 10].

It was investigated that, in cases of partial cement replacement by rice husk ash (RHA) the compressive strength of samples of concrete can be enhanced. The proper proportion of RHA to be added for improvization of the properties of samples lie in the range of 10–30%. The reason for the improvement in mechanical properties could be the fact that it is a pozzolanic material and can make extra C-S-H gel by reacting with Ca $(OH)_2$, obtained during the hydration of cement [18].

It was observed that, when natural aggregate was replaced in brick aggregate by 15%, similar strength is observed. But when 50% replacement of crushed concrete brick was made the compressive strength for 7 days and 28 days are reduced by 14% and 20% respectively. This indifferent behavior may be because of higher absorption capacity and good control of RA grading to a larger container. Due to this property, recycled clay brick aggregate can absorb large quantity of water which may obstruct the hydration of cement at an early age [13–16]. When there is the replacement of coarse aggregate by recycling the waste from floor and wall tiles by 20%, compressive strength remains unaffected. When 100% replacement of floor and wall tiles are done in concrete there is 4.3% and 5.6% decrease as compared to the normal concrete. The compressive strength of NAC (natural aggregate concrete) is similar to RAC (recycled aggregate concrete) made with waste materials as floor and wall tiles [15, 16]. Rather than the various types of waste, the development of compressive strength for recycled aggregate depends on the type of binding mortar as compared to mortar is prepared using 1% vinegar (by weight of cement amount) strength increases by 8.47%, 8.18%, 1.07% after 28 days, 90 days, 180 days, respectively. This may be because of high-strength bonding between the acid attack and hydrated cement paste which may lead to calcium compounds to the calcium salt of the attacking acids. Quality of concrete and solubility of calcium salt depends on the rate of attack [19, 20]. The calcium-containing products in the cement paste react with the acetic acid and produced C-S-H gel and show a positive effect [19]. With the inclusion of marble dust, the basalt concrete and granite concrete shows lower compressive strength for different curing days at 7, 28, and 56 days. It was seen that the water absorption value of river sand and marble sand are almost same [11, 12]. Rather than the different type of marble aggregates, basalt concrete has a higher compressive strength. As compared to basalt and sandstone, river sand has a higher strength [21].

Figures 11.2, 11.3 and 11.4 represent the compressive strength of various industrial wastes as partial replacement of cement concrete [2–8, 11–16].

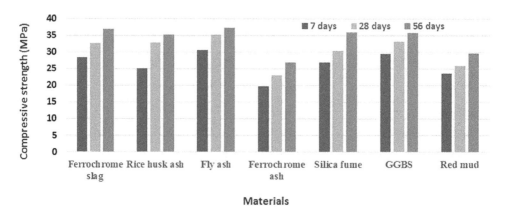

FIGURE 11.2 Compressive strength of concrete for various industrial wastes.

Utilization of Industrial Waste 91

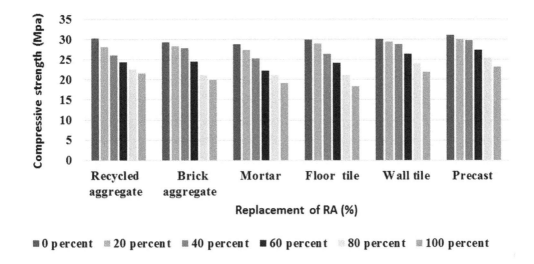

FIGURE 11.3 Compressive strength of concrete for various recycling wastes.

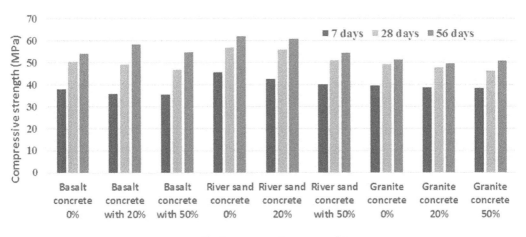

FIGURE 11.4 Compressive strength of concrete for various marble wastes.

11.3.2 Effect of Waste Materials on Tensile Strength and Flexural Strength of Concrete

Another factor that affects the structural performance of concrete is a tensile strength and flexural strength. To demonstrate that the flexural strength of the manufactured concrete increased with the age of concrete. The reason could be the consequence of a major pozzolanic reaction and enhanced interfacial bond bonding between aggregates and paste. It was found that when cement is partially replaced with fly ash, the flexural strength of concrete increases at different curing ages. It was observed that mainly the mechanical strength (tensile and flexural) with different proportions of red mud as a substitution of cement in concrete. The increase in mechanical strength of up to 20% substitution of cement with red mud. Red mud is suitable for ornamental works, or road construction as an embankment landfill [2, 8, 10]. It was observed when Portland cement was partly replaced by 5–10% of GGBS and fly ash by 20%, 40%, and 60%, respectively. The water to cementations materials ratio was maintained at 0.45

for all mixes. The test results proved that the split tensile strength and flexural strength characteristics of concrete improve with the further increments of fly ash and GGBS percentage. It was recorded that suitable proportion of GGBS and fly ash in concrete (9% GGBS and 40% fly ash), however, it was found that upon the further addition of these materials the mechanical strength of the samples does not increase [21–23].

Red mud and hydrated lime were used in concrete as partial replacement material in different percentages to get the pozzolanic properties of red mud. Red mud 0%, 5%, 10%, 15%, 20% and red mud with hydrated lime 5% were replaced to examine compression, flexural, splitting tensile strength test. The result shows that at 15% red mud with hydrated lime gives more flexural strength than controlled concrete and red mud can be used in pavement blocks, embankment landfill, and road surfacing [8]. It was seen that the basalt particles have lamellar geometry shape, which makes weaker zones and prompts be a tensile rupture. This condition may have been enhanced because of the substandard productive intermolecular bond among the cement and basalt particles. Chemical response among basalt and concrete glue brought about a decrease of bond quality [11, 12]. Tensile bond strength among basalt and concrete glue was lesser when contrasted with limestone and quartzite, however, cleavage strength was higher for limestone than that of basalt concrete. The essential reason behind the low concrete glue, bond quality is the manufactured breakdown of feldspars due to their major interaction with the hydrating bond to make soil particles, which swell on retaining water. On the other hand, the substance breakdown of feldspars and other mineral grains on the basalt surface in contact with the solid paste may lessen the surface disagreeableness and weaken the mechanical interlocking effect between the stone surface and the hydration, achieving a weaker bond [11, 12, 24].

It can be noticed that at 15% lime content shows minor differences in performance observed in comparison to normal concrete at the same water-cement ratio. The decrease in strength properties observed with increase in limestone powder content but flexural strength increases gradually. The utilization of waste materials in place of cement enhanced the strength and mechanical properties of concrete. Between 28 and 90 days, flexural strength increased by 13% and 15%, in concrete containing recycling waste and marble dust [11–16] and 2.6–9.3% increase in rebound hammer measurements in comparison to concrete containing limestone aggregate [25]. It was recommended that replacement of cement by 15% of red mud as cement and 30% replacement of quarry dust as fine aggregate gives more compressive strength as well as better flexural characteristics of concrete comparing to the reference sample by use of OPC 43-grade cement [4–10].

Figures 11.5, 11.6 and 11.7 represent the tensile strength and flexural strength for different types of materials [2–8, 11–16].

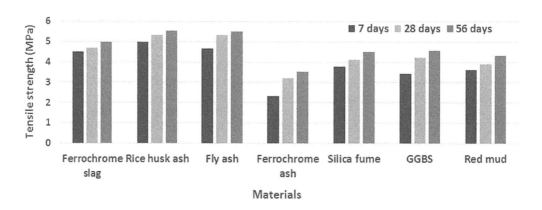

FIGURE 11.5 Tensile strength of concrete for various industrial wastes.

Utilization of Industrial Waste   93

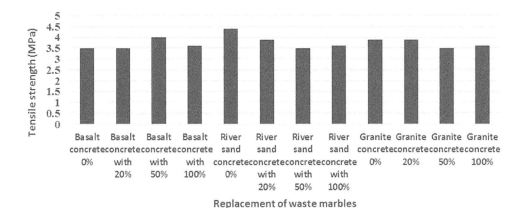

FIGURE 11.6 Tensile strength of concrete for various marble wastes.

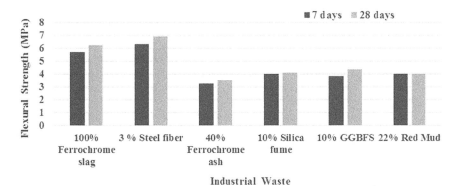

FIGURE 11.7 Flexural strength of concrete for various industrial wastes.

11.4 Conclusion

In this present study, an effort has been made to observe the mechanical properties of industrial wastes as partial replacement of cement in concrete. To investigate the mechanical properties with industrial wastes in cement concrete following conclusions are derived:

- The observed value of the workability, compressive strength, splitting tensile strength and flexural strength indicate that in general, all types of concrete specimens indicate the continuous increase in strength with the development of curing ages.
- Industrial waste materials are suitable as an alternative material of cement and natural coarse aggregate. As a result, a reduction in environmental pollution.
- Performance of all the construction wastes compiled in this study is a suitable replacement of normal concrete. However, construction and demolition waste need suitable treatment before its use.
- The addition of silica fume, fly ash, GGBS independently, along with ferrochrome ash, as partial replacement of cement gives superior mechanical properties in comparison to control concrete. However, the addition of red mud has a marginal influence on the above properties over control concrete.
- The mechanical properties of marble dust, for example, compressive strength, and tensile strength affects the replacement of fine aggregates.
- All the mineral admixtures considered in this study are useful in the preparation of sustainable concrete.

REFERENCES

1. Ajileye, F.V., 2012, "Investigations on micro silica (silica fume) as partial cement replacement in concrete." *Global Journal of Researches in Engineering Civil and Structural Engineering*, 12, 17–24.
2. Acharya, P.K., et al, 2016, "Utilization of ferrochrome waste such as such as ferrochrome ash and ferrochrome slag in concrete manufacturing." *Waste Management & Research*, 34(8), 764–774.
3. Arivalagan, S., 2014, "Sustainable studies on concrete with GGBS as a replacement material in cement" *Jordan Journal of Civil Engineering*, 8(3).
4. Ashok, P., et al, 2011, "Pozzolanic behaviour of compound activated red mud gangue mixture. *Cement and Concrete Research*, 41(3), pp. 270–278.
5. Barbhuiya, S.A., et al, 2009, "Properties of fly ash concrete modified with hydrated lime and silica fume". *Construction and Building Materials*, 23, 3233–3239.
6. Bishetti, P. N., et al., 2014, "Experimental study on utilization of industrial waste in concrete". *International Journal of Technical Research and Applications*, 2(4), e-ISSN: 2320–8163.
7. Corinaldesi, V., et al, 2010, "Characterization of marble powder for his use in mortar and concrete". *Construction and Building Materials*, 113–117.
8. Cachim, P.B., et al, 2009, "Mechanical properties of brick aggregate concrete". *Construction and Building Materials*, 23, 1292–1297.
9. Das, B.B., et al, 2014, *"Characterization of ferrochrome slag as civil engineering material." All India Seminar on Advances in Construction Technology*, Bhubaneswar, Odisha, 54–59.
10. Dang, J., et al, 2018, "Properties of mortar with waste clay bricks as fine aggregate". *Construction and Building Materials*, 166, 898–907.
11. Karaşahin, M., et al, 2007, "Evaluation of marble waste dust in the mixture of asphaltic concrete". *Construction and Building Materials*, 21(3), pp. 616–620.
12. Mathew, A., 2012, "Effect of silica fume on strength and durability parameters of concrete." *International Journal of Engineering Sciences & Emerging Technologies*, 28–35.
13. Mauro, M. T., et al, 2009, "Influence of rice husk ash in mechanical characteristics of concrete." Supplementary Cementing Materials, Paper XII.08, 780–790.
14. Medina, C., et al, 2012, "Microstructure and properties of recycled concretes using ceramic sanitary ware industry waste as coarse aggregate". *Construction and Building Materials*, 112–118.
15. Odler, I., et al, 1987, "Structure and bond strength of cement aggregate interfaces". *MRS Proceedings*, 114, 21–27.
16. Panda, C.R., et al, 2013, "Environmental and technical assessment of ferrochrome slag as concrete aggregate material." *Construction and Building Materials* 49, 262–271.
17. Pavan, K.R., et al 2016, "Experimental study on the effect of cement and sand replacement with red mud and quarry dust in cement concrete pavements." *International Journal on Recent and Innovation Trends in computing a communication*, ISSN: 2321–8169, 4(6).
18. Rashid, K.R., et al, 2017, "Experimental and analytical selection of sustainable recycled concrete with ceramic waste aggregate". *Construction and Building Materials*, 154, 829–840.
19. Sagar, R.R., et al, 2015, "Review on ground granulated blast-furnace slag as a supplementary cementitious material", *International Journal of Computer Applications* (0975–8887) International Conference on Quality Up-gradation in Engineering, Science and Technology.
20. Sakthieswaran, N., et al, 2014, "Compressive strength of concrete containing fly ash, copper slag, silica fume and fibres – prediction." *International Journal of Engineering and Computer Science*, ISSN: 2319-7242, 3(2), 3891–3896.
21. Shetty, M.S., 2017, *"Concrete Technology Theory and Practice"* S Chand & Company Ltd, New Delhi.
22. Silva, D., et al, 2014, "Mechanical properties of structural concrete containing fine aggregates from waste generated by the marble quarrying industry". *Journal of Materials in Civil Engineering*
23. Sowmyashree, T., et al, 2016, "Study on mechanical properties of red mud as a partial replacement of cement with hydrated lime for M40 grade concrete with superplasticizer." *International Journal of Research in Engineering and Technology*, 5(4), ISSN: 2319-1163, ISSN: 2321-7308.
24. Suzuki, M., et al, 2009, "Use of porous ceramic waste aggregates for internal curing of high-performance concrete". *Cement and Concrete Research*, 373–381.
25. Thomas.C.J., et al, 2016, "Structural recycled aggregate concrete made with precast wastes". *Construction and Building Materials*, 114, 536–546.

12

Greenhouse Effect by Investigating an Internal Combustion Engine (IC Engine) Using Argemone Mexicana (Waste Plant) Biodiesel Blends

Akshaya Kumar Rout
KIIT Deemed to be University, Bhubaneswar, India

M.K. Parida
C.V. Raman College of Engineering, Bhubaneswar, India

Mamuni Arya
Radhakrishna Institute of Technology and Engineering, Bhubaneswar, India

CONTENTS

12.1 Introduction 95
12.2 Material and Methods 96
    12.2.1 Oil Preparation Process 96
    12.2.2 Biodiesel Properties 97
    12.2.3 Experimental Procedure 97
12.3 Results and Discussions 98
    12.3.1 Performance Analysis 98
    12.3.2 Emission Analysis 99
12.4 Conclusions 101
References 102

12.1 Introduction

A review of the world energy utilization highlights that a significant fraction of the aggregate energy consumed is obtained from the burning of fossil fuels. These fossil fuels like coal, natural gases and petrochemical sources are largely used in compression ignition (CI) engines, electric power production, transportation, industry, and agriculture. These sources will be consumed shortly due to limited reserve and current usage rates [1]. Among the fossil fuels, fluid petroleum-based powers contribute the maximum owing to their inborn physiochemical and burning properties. The best possible alternative to fossil fuels is biodiesel, which is a clean burning fuel and can be obtained from vegetable oils (edible and non-edible) of plant origin, tree-borne oil seeds, and waste cooking oil. The utilization of edible oil is of great concern being a food material. So, it is defended to utilize non-edible oil for the making of biodiesel. Numerous plant species are present in our country which bear seeds from which we can obtain vegetable oils. It is shocking that, despite their potential, only 6% is utilized. Non-edible oils like Mahua, Jatropha, Karanja, Neem, Polanga, Simarouba, Soapnut, etc., are the different feed stocks available in India [2]. In our country biodiesel can substitute for diesel since huge garbage areas, unutilized open space and country territories are available for cultivation of biodiesel plants. This encouraged recent interest in unconventional sources for petroleum-based fuels.

DOI: 10.1201/9781003217619-13

The Argemone seeds are narcotic and emetic, and they yield (~35%) yellowish brown oil. It is non-edible oil, apparently looking like mustard oil with respect to color, odor, and even specific gravity [3]. Its toxicity is due to the presence of two alkaloids viz. sanguinarine and dihydro-sanguinarine. After removal of toxic alkaloids, the oil might be a good source for biodiesel preparation [Shukla et al, 2005]. Argemone Mexicana belongs to the Papaveracae family, and the entire species belongs to the Mexicana prickle poppy. It is commonly known as shialakanta and satyanashi in India and found on roadsides, wastelands, and fields. The plants have yellow flowers, branching herbs with yellow juice and heights varying between 0.3–0.12m [4]. Refined Argemone oil (RAO), which is alkaloid and gum free, undergoes esterification followed by trans-esterification processes in the presence of acid and alkaline respectively [5] to get Trans-esterified Argemone oil (TEO). The physical and chemical properties of Argemone Mexicana oil blends (B10 and B20) were evaluated, that fall within the range of the American Society for Testing and Materials (ASTM) and European Standards (EN) standard values and are comparable to the conventional diesel properties [6].

The best engine performance was acknowledged for B40 blend at 50% load [7]. Enhanced NO_x with a drop in CO and hydrocarbon (HC) emissions resulted with increase in CR, using methyl ester of waste cooking oil blends. At higher compression ratio, lower values of NHRR, maximum ROPR, higher MFB and longer delay period are obtained for biodiesel blends as compared to diesel. The study revealed that increasing CR had more benefits with biodiesel [8]. The Brake thermal efficiency (BTHE) and engine torque increases for all blends (B10, B20, B30, and B50) as the CR (14–18) increases using wasted cooking oil. The BSFC (Brake Specific Fuel Consumption), CO and HC emission for all blends decreases on the other hand CO_2 and NO_x emission increases as CR increases. The influence of CR (14:1 to 18:1) on combustion and emission parameters at various loads (0–12 kg), using esterified tamanu oil was investigated to find the useful operating CR for biodiesel [9].

In depth review of previous literature and research works suggests that experimental investigations with Argemone Mexicana methyl esters as a fuel in a CI engine are only limited to lower percentage of biodiesel and engine characteristics that are insufficient with respect to the variation in compression ratio. Therefore, the objective is to investigate the effect of CR on engine performance with higher blends.

12.2 Material and Methods

12.2.1 Oil Preparation Process

In the de-gumming process, first the oil was preheated. In 100ml of oil, 25ml of phosphoric acid (H_3PO_4) was added and then heated on a magnetic stirrer up to 1 hour at a temperature of 60°C. Then the oil was kept with 0.1% aqueous NaOH solution and settled in a beaker for 1 day [5]. After that oil was separated and gum particles removed. After separating the oil, we proceed for the next process, i.e., esterification. In the esterification process, preheated oil is taken, in 100ml of oil, 2ml of sulphuric acid and 20ml of methanol is added, and then heated on a magnetic stirrer up to 3 hours at a temperature of 65°C. Then the oil is settled on a separating funnel for 1day. After separating the oil from the separating funnel, the acidic value of the oil is measured: if Free Fatty Acid (FFA) value is less than 2, hence trans-esterification process was carried out. In the trans-esterification process, first oil is preheated, then 100ml of oil is mixed with 20ml of methanol and 0.8gm of Potassium Hydroxide (KOH) and heated on a magnetic stirrer up to 4 hours at a temperature of 55°C. Then the mixture is settled on a separating funnel for 1 day. Here the residue glycerol is separated from the oil.

The purification of biodiesel is carried by a water-washing method. Alcohol is highly soluble in water and oil floats on water; therefore, some amount of water was added in the oil and then the mixture was left to settle on a separating funnel for 1 day. Oil was separated from the lower layer by a separating funnel and washed with water 3–4 times to remove the impurities like acids, methanol. Finally, the biodiesel is heated to nearly 55–60°C to remove the water particles to obtain the resultant trans-esterified oil (TEO) shown in Figure 12.1.

Greenhouse Effect 97

FIGURE 12.1 Trans-esterified Argemone Mexicana oil (AMME).

TABLE 12.1
Properties of Blended Fuel

Sl. No	Properties	Equipment	Method	Diesel	AMME	B20	B40	B60	Limiting Value of AMME
1	Kinematic Viscosity, 40°C (cst)	Cannon-Fenske Viscometer	ASTM D445	3.68	5.07	4.1	4.32	4.58	1.9–6
2	Density at 20°C (kg/m^3)	KEM Density-meter	ASTM D4052	830	868	843	849	858	860–900
3	Calorific values, (kJ/kg)	Bomb Calorimeter	ASTM D240	42800	41500	42540	42280	41980	37830–42050
4	Flash point (°C)	PenskyMartness Apparatus	ASTM D93	65	170	86	107	125	>130°
5	Moisture Content (% wt)		ASTM E-871	0.02	0.2	0.034	0.055	0.070	

12.2.2 Biodiesel Properties

Argemone Mexicana methyl esters (AMME) biodiesel was characterized by its important physical and chemical properties and were measured by the equipment using standard test procedure as per ASTM, and its comparison with standard diesel is summarized in Table 12.1. The properties of fuel blends are found to be within the limiting value. Two step processes reduce specific gravity and viscosity of Argemone Mexicana methyl ester and its diesel blends.

12.2.3 Experimental Procedure

The engine speed was maintained around 1500±3% rpm by controlling the fuel flow by a governor. The load on the engine was varied from no load to full load by adjusting the arm length through an eddy current dynamometer. The engine was tested for standard diesel, B20, B40, B60 and B100 (pure biodiesel) for varying compression ratio from 16–18 by adjusting the stroke volume through a tilted cylinder block arrangement as shown in Figure 12.2. Data were collected after the engine runs for at least 3 minutes for each test fuel. The experiment was repeated three times. The tests were conducted for the above-mentioned fuels at various settings to obtain performance output like BTHE and BSFC. The exhaust gases from the engine were taken through a long tail pipe without increasing the back pressure. For measuring CO, HC, and NO$_x$ emissions, portable multi-gas analyzer (NPM-MGM-1) was used. Pressure data were

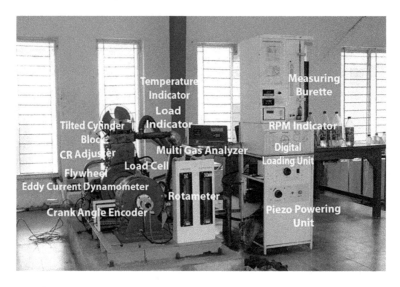

FIGURE 12.2 Multi-fuel VCR engine.

measured of 10 consecutive engine cycles in order to eliminate cyclic variation and average values were taken to analyze and calculate the combustion parameters, such as the peak pressure within the cylinder, NHRR, MFB, ignition delay period and ROPR, etc.

12.3 Results and Discussions

12.3.1 Performance Analysis

Figure 12.3 shows the variation of BSFC with CR for diesel and biodiesel blends at 100% of rated load respectively. At CR 18, BSFC of 0.28, 0.29, 0.31, 0.36 and 0.39 kg/kwh at 100% of rated load was obtained for diesel, B20, B40, B60 and B100 respectively. At a fixed CR, an increase in the concentration

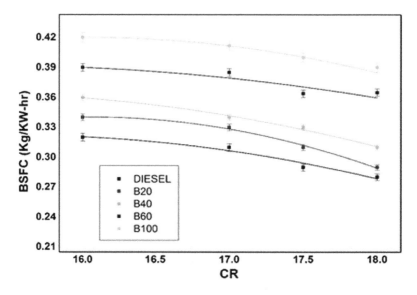

FIGURE 12.3 BSFC vs CR (100% Load).

Greenhouse Effect 99

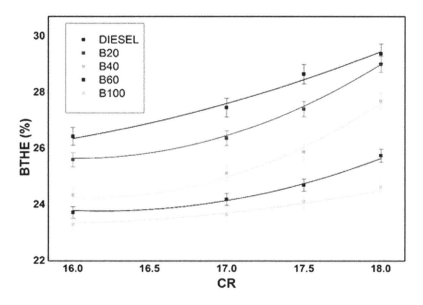

FIGURE 12.4 BTHE vs. CR (100% Load).

of biodiesel in the blend increases BSFC due to the higher value of fuel density and viscosity possessed by the biodiesel. BSFC decreases with increase in CR. It is found to be lower at a compression ratio of 18 for all the test fuels. At CR 16, the highest values of BSFC, i.e., 0.39 kg/kwh and 0.42 kg/kwh, recorded at 100% load for B60 and B100, respectively, may be due to higher value of specific energy consumption at lower compression ratio.

Figure 12.4 shows the variation of the BTHE with CR for different blends at maximum load on the engine. For all the blends, BTHE increases with CR due to the increase in power developed and decrease in heat loss. The BTHE at 12kg load is 29 % for B20 which is at par with diesel at 29.38% because of higher lubrication and oxygen content of lower blends that improves combustion efficiency. For B40, B60 and B100, BTHE of 27.69%, 25.75% and 24.63% are found respectively. BTHE decreases with higher blends due to the lower calorific value and increase in fuel consumption. BTHE of biodiesel is slightly higher at higher CR and lower at low CR, due to inefficient conversion of heat into brake power at lower CR. The result indicates a significant improvement in BTHE for biodiesel in a variable compression ratio (VCR) engine.

12.3.2 Emission Analysis

The comparison of carbon monoxide (CO) emission for standard diesel and biodiesel blends for variation of compression ratios are shown in Figure 12.5 at 100% of rated load. Incomplete oxidation of the air-fuel (A/F) mixture causes CO emission, which depends on A/F ratio, fuel type, atomization rate, and load for a constant speed engine. Better fuel atomization at higher compression ratios leading to proper air fuel mixing and improved combustion efficiency due to oxygen content, causes a decrease in CO emission with increasing compression ratios. The CO emission using diesel, B20, B40, B60 and B100; are found to be 0.021%, 0.018%, 0.029%, 0.037% and 0.026% respectively at CR 18. B60 gives highest CO emission than other blends at all operating CR's. The reason assigned may be higher viscosity and lower calorific value of the fuel with increase in blend ratio, leading to decrease in peak cylinder temperature, resulting in an increase in CO emission.

The comparison of hydrocarbon (HC) emission for standard diesel and biodiesel blends for variation of compression ratios at full engine load are shown in Figure 12.6. Improper combustion and physical properties of fuels influence the HC emissions. Delay period decreases with increase in CR, which improves the combustion process thereby decreases HC emission. The HC emission at peak load, for diesel, B20,

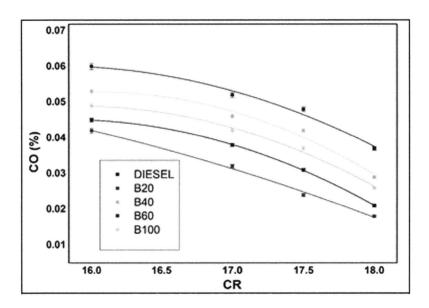

FIGURE 12.5 CO vs. CR (100% Load).

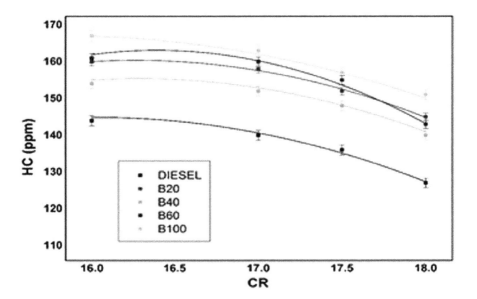

FIGURE 12.6 HC vs. CR (100% Load).

B40, B60 and B100 are 127ppm, 145ppm, 141ppm, 143ppm and 152ppm, respectively. Increases in the percentage of biodiesel in the blend increases the HC emissions. The reason assigned may be due to poor spray formation of biodiesel blends and improper mixing with high-density air, leading to partial combustion of the fuel. Hydrocarbon emission of the blend B40 is found to be better for all compression ratios as compared to other fuels.

Nitrogen oxides (NO_x) formation depends upon CR, pressure and temperature of the inlet air, equivalence ratio and combustion chamber geometry. Figure 12.7 shows comparison of NO_x emissions for diesel

Greenhouse Effect

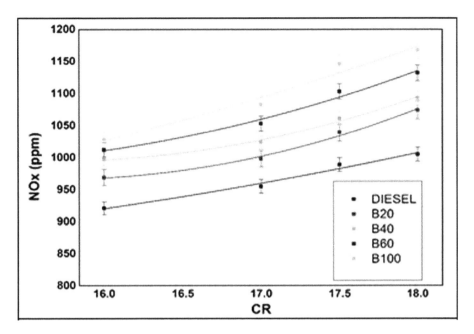

FIGURE 12.7 NOx vs. CR (100% Load).

and biodiesel blends with variation of CR at 100% of rated load. NO_x values of 1.005ppm, 1.074ppm, 1.093ppm, 1.132ppm and 1.168ppm were found for diesel, B20, B40, B60 and B100, respectively, at CR18. The NO_x emission increases with increasing compression ratio for diesel and other blends as ignition delay reduces.

12.4 Conclusions

The following conclusions were observed during the production and testing of Argemone Mexicana Methyl Ester and its diesel blends in a VCR engine.

- The BSFC decreases with increase in CR due to higher value of specific energy consumption at lower CR and increases with increase in biodiesel volume due to the higher value of fuel density and viscosity.
- BTHE increases with the increase in CR due to more power developed and decrease in heat loss and reverse trend obtained with increase in blend ratio owing to the lower heating value and increase in BSFC.
- Decrease in CO emission due to better fuel atomization at higher CR and increase in CO emission with increase in blend ratio because of decrease in peak cylinder temperature is observed.
- Delay period decreases with increase in CR, which improves the combustion process thereby decreases HC emission. HC emissions Increases with increase in biodiesel concentration due to incomplete combustion of the fuel.
- The NO_x emission increases both with increase in CR and concentration of biodiesel in the blend for all the test fuels, due to higher peak pressure and corresponding mean gas temperature within the cylinder.

REFERENCES

1. Srivastava, A., and Prasad, R., (2000). Triglycerides-based diesel fuels. *Renewable Sustainable Energy Reviews*, 4, 111–133.
2. Meher, L. C., Vidya, S. S., Dharmagadda, S., and Naik, N., (2006). Optimization of alkali-catalyzed Transesterification of Pongamia pinnata oil for production of Biodiesel. *Biosource Technology*, 97, 1392–1397.
3. Shukla, A.K., Johar, S.S., and Singh, R.P., (2003). Identification of argemone oil and its simple qualitative detection in mustard oil. *Brassica*, 5, 75–76.
4. Singh, P., Duran, S.K., and Singh, A., (2015). Optimization of biodiesel from Argemone oil with different parameters and performance analysis in CI engine. *International Journal of Research in Engineering and Technology*, 4, 377–386.
5. Pramanik, P., Das, P., and Kim, P.J., (2012). Preparation of biofuel from argemone seed oil by an alternative cost-effective technique. *Fuel*, 91, 81–86.
6. Ariharan, V. N., Gopukumar, S.T., Meenadevi, V. N., Nagendraprasad, P., (2014). Studies on Argemone Mexicana oil for its usage as biodiesel. *International Journal of Pharma and Bio Sciences*, 5, 528–532.
7. Muralidharan, K., and Vasudevan, D., (2011). Performance, emission and combustion characteristics of a variable compression ratio engine using methyl esters of waste cooking oil and diesel blends. *Applied Energy*, 88, 3959–3968.
8. Kassaby, M. E. L., Medhat, M., and Nemitallah, A., (2013). Studying the effect of compression ratio on an engine fueled with waste oil produced biodiesel/diesel fuel. *Alexandria Engineering Journal*, 52, 1–11.
9. Mohanraj, T., and Mohan Kumar, K. M., (2013). Operating characteristics of a variable compression ratio engine using esterified tamanu oil. *International Journal of Green Energy*, 10, 285–301.

13

Fertiliser Plant Phosphogypsum: Potential Applications in Agriculture and Road Construction

Durgasi Hariprasad, Harish Chandra Singh, Pranab Bhattacharyya, and Ranjit Singh Chugh
Paradeep Phosphates Limited, Jagatsinghpur, India

CONTENTS

13.1 Introduction 103
13.2 Materials and Methodology 104
    13.2.1 Experimental Materials 104
    13.2.2 Experimental Methodology 105
13.3 Results and Discussion 107
    13.3.1 Development of Zypmite Product 107
        13.3.1.1 Advantages of Zypmite Product 107
    13.3.2 Phosphogypsum as Road Construction Material 108
        13.3.2.1 Neutralisation of Phosphogypsum 108
13.4 Conclusion 109
Acknowledgement 109
References 109

13.1 Introduction

Phosphatic fertiliser industry generates phosphogypsum and is a by-product from phosphoric acid production plant. Sulphuric acid digestion of phosphate rock generates the co-products of phosphoric acid (H_3PO_4) and phosphogypsum. The raw phosphate (fluorapatite) decomposes with concentrated sulphuric acid at a temperature of 75–80°C. The chemical equation of the reaction is as follows (13.1):

$$Ca_{10}(PO_4)6F_2 + 10\,H_2SO_4 + 20\,H_2O \rightarrow 10\,CaSO_4.2H_2O + 6\,H_3PO_4 + 2\,HF \tag{13.1}$$

Phosphoric acid is mainly used in the production of phosphorus fertilisers: DAP (Diammonium Phosphate) and MAP (Monoammonium Phosphate). For every tonne of P_2O_5 produced as phosphoric acid, 4–6 tonnes dry mass of phosphogypsum are produced depending on rock composition [1]. Worldwide phosphogypsum generation is estimated to be 200–250 million tonnes per year. Phosphogypsum is mainly $CaSO_4 \cdot 2H_2O$ but also contains impurities such as H_3PO_4, $Ca(H_2PO_4).2H_2O$, $CaHPO_4.2H_2O$ and $Ca_3(PO_4)_2$, residual acids, fluorides (NaF, Na_2SiF_6, Na_3AlF_6, Na_3FeF_6 and CaF_2), sulphate ions, traces of heavy metals, adhered to the surface of the gypsum [2]. The nature and characteristics of the resulting phosphogypsum are strongly influenced by the phosphate ore composition and quality. Phosphogypsum is generated, and this product, by not having direct and continuous use, and also by being currently stored in landfills exposed to the weather, becomes one of the biggest environmental problems for fertiliser producers. Phosphogypsum, discharged into the sea, watercourses or in wilderness stocks pit, contains toxic elements harmful to ecosystems and human health, including heavy metals and radionuclides, and there is

DOI: 10.1201/9781003217619-14

therefore a concern regarding environmental impacts. The concentrations of these elements vary between the regions and the processes used, all of which require particular and specific follow-up after the release of phosphogypsum and during its use. Phosphogypsum is used in agriculture for soil amendment or as fertiliser, as well as in the brick and cement industry, and in road construction.

Phosphogypsum, like gypsum from any source, is an almost ideal agricultural source of both calcium and sulphur. In areas where phosphogypsum is available, it has the advantage of being the most economical of all types of gypsum for agricultural use. Since it is slowly soluble, phosphogypsum remains available to the plant over long periods. Unlike lime which makes the soil more alkaline, phosphogypsum is neutral in its soil reactions. As a calcium source, gypsum is the material of choice for peanut farmers and, while it is a small market, gypsum used to supply calcium to ferns raised for the floral market allows the soil to remain acidic for optimum growth. In other words, gypsum is a hard-to-beat source of calcium in agriculture. Agricultural literature is reporting ever increasing instances of soil sulphur deficiencies that adversely affect crop yields. With increased emphasis on removing sulphur gases from the air and the use of high analysis fertilisers that contain little or no sulphur, the previous most common methods of replenishing soil sulphur no longer exist.

We focus on the application of phosphogypsum in agriculture. Phosphogypsum can be used for improvement of both alkaline and acidic soils by: (i) slowing down soil degradation; (ii) increasing soil permeability, which can reduce the surface runoff and soil erosion; (iii) boosting the availability of the organic matter and phosphorus in the soil; (iv) improving salt washing and alkali leaching, preserving soil moisture, and protecting against drought and water logging; and (v) effectively reducing ammonia volatisation in composting process, and improving nitrogen use efficiency.

The overdosing of chemical fertilisers and deficiency of micronutrients in the soil has stopped the rise in crop yield. The soil is no more neutral and requires neutralising agents like gypsum to be applied. The use of micronutrients like Zn, B, Fe and S with nitrogen, phosphorus and potash (NPK) will make significant Improvement in the crop yield. Zypmite containing nutrients for high yield of crops with good nutrient value. Zypmite project is economically attractive for farmers, as a part of waste product gypsum can be used to manufacture value added product like Zypmite Plus.

Non replenishment of the micronutrients to the soil is responsible for continuous decline in the crop response ratio as the Indian soils have been depleted with vital micronutrients required for crop production. Sixteen nutrients are required in the soils for crop production. Of late, significant decline in micronutrients like S, Zn, Mg, B, Fe, Mn has been observed. Zypmite, a granulated phosphor-gypsum is used as plant nutrient providing much needed sulphur (S) and calcium (Ca) to the soil. PPL's Zypmite product will make it logistic friendly and easy to broadcast the product.

In view of characteristics of phosphogypsum (more than 95% is $CaSO_4.2H_2O$) and its attractive economic potential, as well as continuously increasing concerns about environmental pollution, nowadays there is great interest in using phosphogypsum as an alternative raw material for many applications. Phosphogypsum has been used in the cement industry as a setting regulator in place of natural gypsum [3] and in the gypsum industry to make gypsum plaster. Phosphogypsum also has been used as agricultural fertilisers or soil stabilisation amendments [4] and as a fly ash-lime reaction activator with wide application in the manufacturing of building materials [6]. Studies have shown that phosphates and fluorides delay the setting time and reduce the early strength development of cement and has been found that a relatively high level of affects several aspects of gypsum crystal formation [7]. Thus, processes based on washing, drying, and chemical and thermal extraction have been recommended by numerous researchers to make use of phosphogypsum.

13.2 Materials and Methodology

13.2.1 Experimental Materials

In the present work phospogypsum (Figure 13.1) taken for the Zypmite product development. The chemical and physical analysis and mineralogical phases of phosphogypsum generated from phosphoric acid plant of M/s: Paradeep Phosphates Ltd is given in Table 13.1.

Fertiliser Plant Phosphogypsum

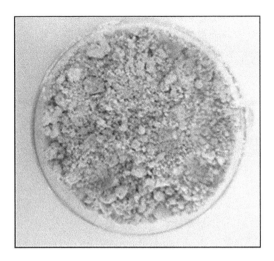

FIGURE 13.1 Appearance of phospogypsum with moisture.

TABLE 13.1
Chemical and Physical Analysis of Phosphogypsum

S.No.	Parameters	% (wt/wt)
1	Free Moisture	15.400
2	Purity as $CaSO_4 \cdot 2H_2O$	95.020
3	Silica & Insoluble	1.180
4	Fe_2O_3	0.025
5	Al_2O_3	0.240
6	CaO	31.970
7	MgO	0.030
8	Na_2O	0.200
9	K_2O	0.020
10	Total P_2O_5	0.710
11	Water Soluble P_2O_5	0.170
12	SO_4^{2-}	53.000
13	Fluoride	0.820
14	Chloride	0.006
15	Particle Size (–100 mesh)	96.200
16	pH of 5% solution	4.2–5.5
17	Bulk Density gm/cc	1.16

Phosphogypsum chemical and physical analysis were carried by classical methods shows majority of the matrix is moisture 15.4%, calcium oxide 31.97%, sulphate 53% and silica, and water insolubles (1.37%) and the rest of the minor constituents are oxides of Na, Al, P, Cl, K, Ti, Fe, etc. (< 1%). From the XRD analysis Regaku XRD, model no Ultima –IV, Cu target the phase is identified as $CaSO_4 \cdot 2H_2O$ (JCPDS card no 00-033-0311) Figure 13.2.

The particle size of phosphogypsum is very fine and the various fraction shown in Table 13.2.

13.2.2 Experimental Methodology

The product Zypmite (Figure 13.3) is innovated using Phosphogypsum mixed along with an aluminium-silicate-based binder and synthetic plant nutrients in a paddle mixer. Thoroughly mixed composite from the mixer comes to the granulator drum. Adequate quantity of water and granulating aid is sprayed

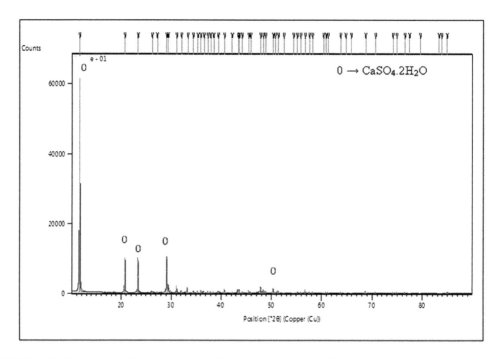

FIGURE 13.2 X-ray powder diffraction pattern of Phospogypsum with phase identification.

TABLE 13.2

Phosphogypsum Particle Size Analysis

Sl No.	Sieve Size, mm	% Passing
1	4.75	100
2	2.36	100
3	1.8	85
4	0.06	79
5	0.042	77
6	0.030	73
7	0.015	66
8	0.075	51
9	0.002	0

on the rolling down particles. The fine spray wets the surface of the particles where fine powder get deposited and the small particles achieve the desired size of granules by an agglomeration process. The granule size depends upon the speed and size of the granulator, amount of water sprayed and the residence time of granules. The granules formed in the granulator are fed to a rotary drier followed by product cooler, screeing and packing. The final 2–4mm granule achieved 0.5–06.5 kg force crushing strength. However, the crushing strength of the Zypmite product, depends on the type of nutrient's additives, moisture level, exposure temperature and formulation.

Fertiliser Plant Phosphogypsum

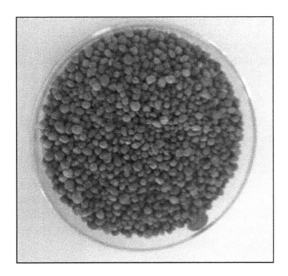

FIGURE 13.3 R&D developed Zypmite product from Phosphogypsum.

13.3 Results and Discussion

13.3.1 Development of Zypmite Product

The laboratory developed Zypmite product physical and chemical analysis results are shown in Table 13.3.

13.3.1.1 Advantages of Zypmite Product

Zypmite is a micronutrient mixture containing Sulphur, Zinc, Boron, Calcium and Magnesium. Zypmite fortified helps improve soil fertility, increases the intake of NPK fertilisers and improves quality of yield. Moreover, it improves plant health, improves soil condition, increases oil content in oil seeds, allows better fruit development, improves growth of plants and roots, improves metabolic activity and prevents fruit cracking. Zypmite is a cheapest sulphur source containing product.

TABLE 13.3

Analysis Results of Final Zypmite Product

Sl. No.	Product Parameters	Unit	Analysis Result
1	Moisture	% w/w	4.5
2	Sulphur as S	% w/w	13.25–14.7
	Calcium as Ca	% w/w	20.0
3	Purity as $CaSO_4 \cdot 2H_2O$	% w/w	71.20
4	Crushing Strength	kg force	0.55–0.65
	Particle size		
	+4.0mm		8
	−4.0mm to + 3.35mm		12
5	−3.35mm to + 2.0mm	% w/w	65
	−2.0mm to + 1.0mm		15
	−1mm		0
6	Boron as B	% w/w	0.11
7	Zinc as Zn	% w/w	1.0–1.8
8	Magnesium as Mg	% w/w	0.3–0.45

13.3.2 Phosphogypsum as Road Construction Material

Initially the phosphogypsum is neutralised with hydrated lime further if it is applied as road construction material. The detailed work is discussed in the following section.

13.3.2.1 Neutralisation of Phosphogypsum

The above-mentioned Table 31.1 shows specifications for phosphogypsum taken for the exploration and feasibility study. The neutralisation with lime is suggested by CSIR-CRRI, New Delhi to modify the quality of Phosphogypsum so as to remove the impurities or reduce their level. It is clear that; neutralisation of phosphogypsum takes care of the free acid impurities by making them inactive. Based upon the cumulative results, it is found that, about 2.5~3.0% of total volume of gypsum would be the volume of lime required to be added for complete and proper neutralisation, which makes the water soluble P_2O_5 in untraceable limits. In addition, PPL engaged the Central Road Research institute (CRRI) to assess the suitability of using neutralised gypsum as a road construction material. The broad scope of work carried out by CSIR-CRRI in collaboration with PPL Ltd is as follows:

a. Characterisation of phosphogypsum
b. Characterisation of soil
c. Use of phosphogypsum in road embankment and subgrade
d. Use of phosphogypsum in concrete road

Study was conducted by collecting the sample locally from the plant. In order to use phosphogypsum, several proportions of mixes were tried to find out the strength. In order to assess the efficiency of phosphogypsum for the purpose of stabilisation, soil has been stabilised with phosphogypsum and phosphogypsum+soil, respectively. Strength gain was determined in terms of Unconfined Compressive strength (UCS). Durability tests were also conducted to assess the performance of stabilised soils when subjected to wetting and drying for simulating water logging and flooding situations at the site.

Based on laboratory studies, it is concluded that phosphogypsum, as such, can be used as a fill material and in subgrade/subbase layer of a road pavement. Local soil stabilised with 20% phosphogypsum can be used a subgrade/capping layer. Local soil stabilised with 20% phosphogypsum and 2% or 4% lime can also be used as subgrade/capping layer.

However, CRRI recommended proceeding with small-scale application before proceeding to large-scale applications. As per guidelines of CSIR-Central Road research institute (CSIR-CRRI), PPL proposes to carry out trials in their plant premises at Paradeep, using phosphogypsum for road construction. In line with studies at CRRI above, a location has been finalised to construct the experimental road along

FIGURE 13.4 Trail experimental road using neutral phosphogypsum.

Fertiliser Plant Phosphogypsum 109

the existing road to a gypsum pond so that the experimental road can be monitored over a period of time due to movement of heavy duty vehicles used for transportation of gypsum from gypsum pond length of 500m. This has been considered and constructed for trials at PPL Gypsum Pond premises (Figure 13.4). The durability of the trials road-tested with heavy vehicle movement over it lasts for 2 monsoon seasons.

13.4 Conclusion

With rapid growth in the agriculture production in India, high-concentration phosphate in compound fertilisers, phospogypsum production is increasing every year. Phosphogypsum utilisation and environmentally friendly treatment are becoming an increasing concern for sustainable development. Phosphogypsum is discussed in this work in order to provide the scientific support for the effective utilisation. Comprehensive utilisation of phosphogypsum may greatly help in saving other precious resources and increasing economic benefits.

1. Zypmite is a micronutrient mixture product containing sulphur, zinc, boron, calcium and magnesium. Zypmite fortified helps improve soil fertility, increases the intake of NPK fertilisers and improves quality of yield.
2. Moreover, it improves plant health, improves soil condition, increases oil content in oil seeds, better fruit development, improves growth of plants and roots, metabolic activity and prevents fruit cracking. Zypmite is a cheapest sulphur source containing product.
3. Based on laboratory studies, it is concluded that phosphogypsum, as such, can be used as a fill material, and in subgrade/sub base layers of a road pavement.
4. Local soil stabilised with 20% phosphogypsum can be used a subgrade/capping layer. Local soil stabilised with 20% phosphogypsum and 2% or 4% lime can also be used as subgrade/capping layer.
5. As per guidelines of CSIR-Central Road research institute (CSIR-CRRI), PPL proposes to carry out trials in their plant premises at Paradeep Township, using phosphogypsum for road construction.
6. An experimental road along the existing road to a gypsum pond, so that the experimental road can be monitored over a period of time due to movement of heavy duty vehicles used for transportation of gypsum from gypsum pond length of 500m has been considered and constructed for trials at PPL Gypsum pond premises.
7. The durability of the trials road tested with heavy vehicle movement over it lasts for 2 monsoon seasons.

ACKNOWLEDGEMENT

We are thankful to Adventz group Chairman Mr Saroj Kumar Poddar, and Mr N. Suresh Krishnan (MD), PPL, and Paradeep for their kind support in this work. We appreciate CSIR-IMMT, Bhubaneswar for the characterisation of gypsum and products samples. One of the authors, Dr D Hariprasad, PhD (Manager R&D), especially thanks Shri Pranab Kumar Bhattacharyya, CGM (Operations) and Shri Ranjit Chugh (COO) for their constant support, encouragement, and motivation on innovation of new products and technology developments. We express gratitude to Department of Scientific and Industrial Research (DSIR) for encouraging the indigenous technology promotion, development, and utilisation.

REFERENCES

1. Wan, B. Hu, W.H. *Science and Technology of West China*, 25, 5–7, 2011.
2. Rutherford, P.M., Dudas, M.J., Arocena, J.M., Heterogeneous distribution of radionuclides, barium and strontium in phosphogypsum by-product. *Science of the Total Environment* 180 (3), 201–209, 1996.

3. Akın, A.I., Yesim, S., Utilization of weathered phosphogypsum as set retarder in Portland cement. *Cement and Concrete Research* 34 (4), 677–680, 2004.
4. Papastefanou, C., Stoulos, S., Ioannidou, A., Manolopoulou, M, The application of phosphogypsum in agriculture and the radiological impact. *Journal of Environmental Radioactivity* 89 (2), 188–198, 2006.
5. Degirmenci, N., Okucu, A., Turabi, A., Application of phosphogypsum in soil stabilization. *Building and Environment* 42 (9), 3393–3398, 2007.
6. Weiguo, S., Mingkai, Z., Qinglin, Z., Study on lime–fly ash–phosphogypsum binder. *Construction and Building Materials* 21 (7), 1480–1485, 2007.
7. Koopman, C., Purification of gypsum from the phosphoric acid production by recrystallization with simultaneous extraction. PhD thesis, TU Delft, 2001.

Part II

Waste Utilization and Soil Stabilization

14

Bearing Capacity of Reinforced Soil on Varying Footing Size

Bandita Paikaroy
KIIT, Deemed to be University, Bhubaneswar, India

Sarat Kumar Das
I.I.T (ISM), Dhanbad, India

Benu Gopal Mohapatra
KIIT, Deemed to be University, Bhubaneswar, India

CONTENTS

14.1 Introduction 113
14.2 Model Footing Test 114
14.3 Model Test Results and Discussion 116
14.4 Conclusion 117
References 117

14.1 Introduction

The bearing capacity equation derived by Terzaghi (1943) is used to find the (UBC) ultimate bearing capacity (q_{ult}) of soil. Terzaghi (1943) expressed the bearing capacity equation for a rough strip foundation founded at depth 'D' below the surface. The well-known equation is presented as

$$Q_{ult} = c\,N_c + \gamma\,D\,N_q + 0.5\,\gamma\,B\,N_\gamma$$

The ultimate bearing capacity depends on the unit weight of soil, width of footing, cohession and internal friction of soil and the bearing capacity factor. For the special case, footing placed on surface of soil mass is expressed as

$$Q_{ult} = 0.5\,\gamma\,B\,N_\gamma$$

The bearing capacity factors used in equation are depending on the angle of internal friction affected by the relative density of the material present below foundation. Many researchers have studied the effect of footing size on bearing capacity analysis associated with N_γ (Berry, 1935, De Beer, 1963, Dewaiker & Mohapatra, 2003, Cerato & Lutenegger, 2006). Also, the bearing capacity is dependent on the type and laying pattern of reinforcement below footing. (Adam & Collin, 1997, Dash et al., 2001, Bera et al., 2005, Mosallanezhad and Hataf, 2008). Many experimental as well as numerical and analytical studies have been done on reinforced foundation on different condition taking sand as a foundation material, but limited studies are there on bearing capacity using industrial waste in foundation (Bai et al., 2013). Exploring the use in stabilization process has been investigated by Gupta et al., 2002, Gulsah, 2004, Soosan et al., 2001, Soosan et al., 2005 and Naganathan et al., 2012.

DOI: 10.1201/9781003217619-16

113

This paper presents and discusses the results of square footing of size 15cm and 18cm on loose density of the granular material (crusher dust). It focuses the result of variation of footing size on UBC and N_γ. Crusher dust, an industrial waste, is produced from rubble crusher unit during the formation of aggregates from the rubble in huge quantities. Its accumulation on the same site is producing a health hazard and air pollution to great extent. Like other industrial waste, it can be better implemented below foundation, fills and pavements, in large quantity, in turn, this adverse effect can be minimized. The present work is performed on crusher dust. Its behavior is like a poorly graded sand. So, without use of sand, this waste has been filled in the model test tank contemplated with geogrid as reinforcement. Keeping in mind of environmental pollution to a large extent by the extraction of industrial wastes, i,e., crusher dust, a matter of worry to use it in better way for better prospects. In the present study, it has been used as a foundation material below surface square footings of different sizes at loose density state.

14.2 Model Footing Test

Model footing tests were performed in a 1.6m × 1.0m × 1.2m steel tank with a perspex sheet at the front, Figure 14.1. The perspex sheet gives a good visualization of the layered material filled in the test tank with uniform density throughout, from bottom to the top. All the tests were conducted at the desired densities of $D_r = 29\%$. The footing was resting on the crusher dust surface ($D_f = 0$). The dimensions of square footings were 0.15m and 0.18m. The dust was poured into the tank from a height of 0.15m to the target density. The load cell was attached to the square plate through an attachment rod. Two LVDTs (Linear Variable Differential Transducers) were attached vertically on the plate in diagonal positions for measuring the settlement during the test, and the results were obtained from the load-settlement curve. The load was increased at a rate of 0.008kN/sec for each test. The crusher dust used as the foundation material is exhibited as a sandy type of soil obtained from the grain size distribution presented in Figure 14.2. The coefficient of uniformity and coefficient of curvature are found as 10.55 and 0.47, respectively with maximum and minimum void ratios as 0.591 and 0.323 respectively.

Physical properties of crusher dust obtained following ASTM D4253-16 indicate a poorly graded soil. The crusher dust was poured into the tank following a rainfall technique. The reinforcements used for these experiments were of a geogrid. Its tensile strength was found to be 4.3kN/m (ASTM D4595-17). The top layer depth 'u' and the successive depth 'h' of reinforcement geogrid was considered as 0.25B and 0.5B, where 'B' is the footing size. After reaching to the required depth, the surface was leveled and the geogrid was placed centrally. This process was continued till the top surface was reached. The layout of geogrid over the bed of crusher dust is presented in Figure 14.3.

Footing was placed over the crusher dust surface after filling the material up to top of the tank at required level as shown in Figure 14.4.

In this method of test, a double tangent method was used to find the ultimate bearing capacity of footing.

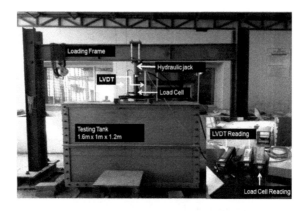

FIGURE 14.1 Model footing tests were performed in a 1.6m × 1.0m × 1.2m steel tank with a perspex sheet at the front.

Bearing Capacity of Reinforced Soil 115

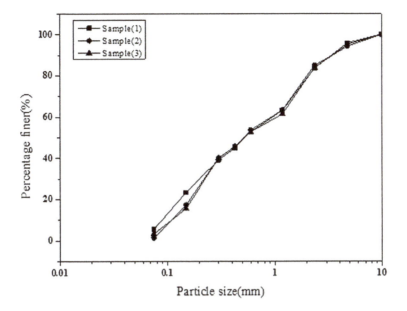

FIGURE 14.2 Layout of geogrid on crusher bed.

FIGURE 14.3 Grain size distribution of crusher dust.

FIGURE 14.4 Model test plate at failure.

14.3 Model Test Results and Discussion

Geogrid as reinforcement was used in 3-layer basis considered as N = 1 to N = 3 whereas 'N' stands for number of reinforcement layer. Typical load-settlement curves obtained for square footing of B = 0.18m at varying layers of geogrid are presented in Figure 14.5.

Table 14.1 presents the results of the UBC of both the square model footings at loose density of crusher dust. At same density with increase in reinforcements, the UBC is increasing with respect to unreinforced condition. The larger the footing size (B), the higher is the q_{ult}. This is consistent with Terzaghi's (1943) bearing capacity equation which shows that for surface footing on sandy soil, 'B' affects the increase/decrease in result. The increase in footing size exhibiting 8.5% increase in q_{ult} at a Relative Density (RD) of 29%. The results of UBC for both the footings are tabulated in Table 14.1.

The N_γ (bearing capacity factor) has been back calculated and plotted on the graph shown in Figure 14.6. It is increasing with reinforcement layers and decreasing with footing size. This trend is true for any type of geosynthetic and any density variation. Cerato and Lutenegger (2006) obtained the same trend for increase in density and varying the footing size, taking square and circular footings.

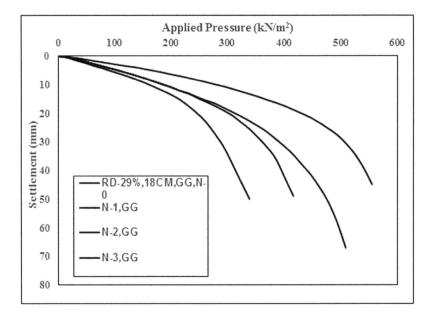

FIGURE 14.5 Load-settlement curve for square 0.18m footing at varying geogrid layers.

TABLE 14.1

Results of Ultimate Bearing Capacity at Both Densities for Two Footings

Relative Density (RD)%	Footing Type	Size of Footing (B)	Reinforcement	No. of Reinforcement Layers	Bearing Capacity (q_{ult}) kN/m²
29	Square	15cm	Geogrid	N=0	235
				N=1	290
				N=2	410
				N=3	440
	Square	18cm		N=0	255
				N=1	337
				N=2	430
				N=3	485

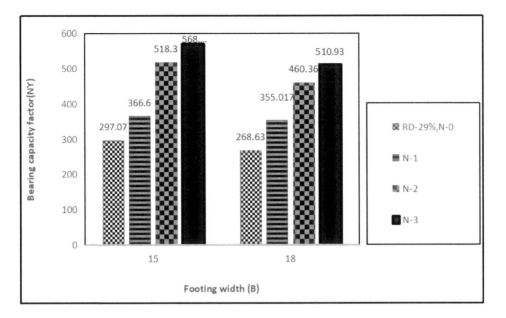

FIGURE 14.6 N_γ comparison of square footing.

Figure 14.6 shows the trend of increase in bearing capacity factor (N_γ) for both footings. It is showing 10.5% rise at RD of 29% at an unreinforced state and at 3.2%, 12.58%, and 11.3% for single, double and three layers of reinforcement. For the first to second layer, appreciable increase is observed, whereas for the second to third layer, minimal increase is obtained. At same density, the 0.15m(B) footing is producing a higher N_γ value than the 0.18m(B) size of footing.

14.4 Conclusion

Results of model footing tests show the following bearing capacity features.

1. Dependent on relative density, number of reinforcement layers, addition of reinforcement to the foundation material.
2. Addition of reinforcement gives appreciable increase in bearing capacity compared to unreinforced condition of foundation material. It is increasing with increase in footing size.
3. The bearing capacity factor((N_γ) in increasing with increase in number of geogrid layers and decreasing with increase in footing size. At RD = 29%, it is showing a 10% increase.
4. Further research, taking higher size of footing with higher density, will produced more pronounced result.

REFERENCES

Adam, M.T., Collin, J.G., 1997. Large model spread footing load tests on geosynthetic reinforced soil foundation. *Journal of Geotechnical Engineering*, ASCE 123(1), 66–72, doi: 10.1061/(ASCE)1090-0241(1997)123:1(66).

ASTM, 2016. D4253-16, Standard test methods for maximum index density and unit weight of soils using a vibratory table, ASTM International, West Conshohocken, PA.

ASTM, 2017. D4595-17, Standard test method for tensile properties of geotextiles by the wide-width strip method, ASTM International, West Conshohocken, PA.

Bai X-H., et al., 2013. Bearing capacity of square footing supported by a geobelt-reinforced crushed stone cushion on soft soil. *Geotextiles and Geomembranes*, 38, 37–42. doi: 10.1016/j.geotexmem.2013.04.004.

Bera, A.K., et al, 2005. Regression model for bearing capacity of a square footing on reinforced pond ash. *Geotextiles and Geomembranes* 23(2), 261–285, doi: 10.1016/j.geotexmem.2004.09.002.

Berry, D.S., 1935. Stability of granular mixture. *Proc., 38th Annual Meeting*, ASTM, PHiladelphia, 35, 491–507.

Cerato, A.B., Lutenegger, A.J., 2006. Bearing capacity of square footing and circular footings on a finite layer of granular soil underlain by aa rigid base. *Journal of Geotechnical and Geoenviromental Engineering*, 132(11), 1496–1501.

Dash, S.K., et al, 2001. Strip footing on geocell reinforced sand beds with additional planar reinforcement. *Geotextiles and Geomembranes*, 19, 529–538, DOI: 10.1016/S0266-1144(01)00022-X.

De Beer, E.E., 1963. The scale effect in the transportation of the results of deep-sounding tests on the ultimate bearing capacity of piles and caisson foundation. *Geotechnique,* 13(1), 39–75.

Dewaiker, D.M., Mohapatra, B.G., 2003. Computation of bearing capacity factor N_γ - Terzaghi's mechanism. *International Journal of Geomechanics*, 3(1), 123–128.

Gulsah., 2004. Stabilization of expansive soils using aggregate waste, rock powder and lime, a Master of Science thesis, submitted to the graduate school of natural and applied sciences of the Middle East technical university.

Gupta, A.K., et al, 2002. Stabilization of black cotton soil using crusher dust –A waste product of Bundelkhand region, *Proceedings of Indian Geotechnical Conference*, Allahabad, 308–311.

Mosallanezhad, M., Hataf, N. (2008). Experimental study of bearing capacity of granular soils, reinforced with innovative grid-anchor system. *Geotechnical and Geological Eng,* 26(3): 299–312.

Naganathan, S., et al, 2012. Properties of controlled low-strength material made using industrial waste incineration bottom ash and quarry dust. *Materials & Design*, 33, 56–63. doi: 10.1016/j.matdes.2011.07.014.

Soosan, T.G., et al, 2001. *Use of Crusher dust in embankment and highway construction. Proceedings of Indian Geo-Technical Conference*, December, Indore, 274–277.

Soosan, T.G., et al, 2005. Utilization of quarry dust to improve the geotechnical properties of soils in highway construction. *Geotechnical Testing Journal,* 28(4), doi: 10.1520/GTJ11768.

Terzaghi, K., 1943. Bearing Capacity, *Theoretical soil mechanics*, Chap. 8, 118–143.

15

Improvement of Properties of an Expansive Soil with Induction of Bacteria

Suresh Gunji, C. H. Sudha Rani, C. H. Paramageetham, Yugandhar Gundlapalli, and Basha Sreenivasulu
Sri Venkateswara University, Tirupati, India

CONTENTS

15.1 Introduction 119
15.2 Experiment Investigations 120
    15.2.1 Materials 120
        15.2.1.1 Soil 120
        15.2.1.2 Bacteria 120
15.3 Tests Conducted 121
    15.3.1 Atterberg Limits 121
    15.3.2 Unconfined Compression Strength Test 121
    15.3.3 Soil Surface Morphology 121
    15.3.4 pH Value 121
15.4 Test Results 121
15.5 Plasticity Characteristics 121
15.6 Strength 124
15.7 pH Value 124
15.8 Micro Studies 126
15.9 Conclusions 127
References 127

15.1 Introduction

Expansive soils also known as Black Cotton Soils exhibit the characteristic feature of swelling and shrinking with changes in moisture content, causing damages to pavements, runways, and building foundations, which are founded on these soils (Chen 1988, Nelson & Miller 1992). Chemical and mechanical stabilization methods have been applied with varying success rates to stabilize expansive soils. Cement, lime, fly ash and granulated blast furnace slag have been used for the treatment of expansive soils for decades (Obuzor et al. 2011). Nowadays, various soil stabilization techniques are being practiced. In recent years, the use of Microbial Induced Calcium Carbonate Precipitation (MICP) has become a popular soil stabilization technique. It is a new and environmentally eco-friendly method (Dejong et al. 2006, Karol 2003). This method has an advantage over conventional chemical treatments; it can be non–toxic and environmentally favorable and have a limited injection distance. It is also cost-effective in comparison to chemical treatments (Ivanov & Chu 2008). However, limited studies have been established related to the implementation of MICP in expansive soils. The microbial population is a significant obstacle to the application of MICP in expansive clays. The major role of bacteria in the Calcite Induced Precipitation process has been to produce an alkaline environment throughout different physiological actions (Douglas & Beveridge 1998; Dejong et al. 2010; Benini et al. 1999; Van Paassen et al. 2010a; Hamdan et al. 2011;

DOI: 10.1201/9781003217619-17

Warthmann et al. 2000; Roden et al. 2002; Burbank et al. 2011 and Basha et al. 2018). The method of Calcium Carbonate Precipitation by urease-producing bacteria consists of urea hydrolysis and Calcium Carbonate Precipitation (Hammes & Verstraete 2002, Burbank et al. 2013). Burbank et al. (2011) found that microorganisms capable of hydrolyzing urea to carbon-dioxide and ammonia are widespread in soils. According to Lloyd & Sheaffe (1973), 17% and 30% of cultivable aerophilic, microaerophilic, and anaerobic microorganisms are capable of hydrolyzing urea. In MICP, one Mole (1M) of urea $(NH_2)_2CO$ is hydrolyzed to two Moles (2M) of NH_4^+ (ammonia) and one Mole (1M) of CO_3^{2-} by the microbial enzyme urease, represented as $(NH_2)2CO +2H_2O_2$ producing $NH_4^+ + CO_3^{2-}$. In the absence of calcium ions, CO_3^{2-} gets precipitated as $CaCO_3$ $(Ca^{2+} + CO_3^{2-} \rightarrow CaCO_3)$. The development of NH_4^+ enhances the soil pH (8.5) and more significantly, the rate of calcium carbonate precipitation (Hammes & Verstraete 2002). MICP creates a connection between soil grains cementing soil grains together (DeJong et al. 2006). *Bacillus pasteurii*, a highly urease-active bacteria, plays an important role in calcium carbonate or calcite $(CaCO_3)$ precipitation (Bangetal et al. 2001; Dejong et al. 2006; Dejong et al. 2010). The *B. pasteurii* has been reclassified as Sporosarcinapasteurii (Mitchell & Ferris 2006). MICP varies with the procedural parameters, including the flow rates, the flow direction, and the formulations of the biological and chemical amendments. The physical, chemical, and biological properties are essential to the performance of MICP, like shear wave velocity, permeability, calcium carbonate content, calcium, ammonium, urea, and bacterial density. MICP can be used in a wide range of soil types and from conditions of freshwater to full-strength seawater and various degrees of saturation, with better results for low degrees of saturation (Mortensen et al. 2011; Cheng et al. 2013). It has generated up to 6% by weight calcium carbonate within several treatment days (Burbank et al. 2011; Chu et al. 2012), whereas denitrification based MICP needed several months, up to a year, to obtain an average 1% and 3% by weight (van Paassen et al. 2010a; Van et al. 2011, O'Donnell 2016). Chittoori and Neupane (2018) analyzed the application of bioremediation to reduce expansive soil swelling and noted that low to medium plastic soils can be effectively treated with MICP by bacteria. Induced bacterial is produces calcite precipitation in expansive soil to reduce swelling, shrinkage and may not have any allowable limits; it depends on the depth of treatment (Cheng & Cord-Ruwisch, 2012). Microbially Induced Calcium Precipitation process is effectively used in expansive soil stabilization (Burbank. et al. 2018; Chittoori et al. 2018; Pusadkar et al. 2019). MICP treatment is a simple, biologically induced process for in-situ cementation and improvement of mechanical soil properties (Montoya and DeJong 2015). MICP can be used for a range of applications including improvement of soil strength and hardness, improvement and remediation of concrete, alternative building materials and environmental remediation (Li et al. 2016). MICP is a sustainable biological ground improvement technique that is capable of altering and improving soil mechanical and geotechnical engineering properties (Cheng & Shahin 2016; Liang Cheng et al. 2016). The use of MICP to stabilized expansive soil through bacteria is still a hypothesis. This analysis is a first step toward assessing the viability of this as hypothesis and obtaining a deeper understanding of the challenges of stabilizing expansive soil (Martinez et al. 2013). MICP could be used for expansive soil treatments, according to these studies, and further research is being done to evaluate thresholds at which MICP could be used effectively in expansive soil treatments.

15.2 Experiment Investigations

15.2.1 Materials

15.2.1.1 Soil

An expansive soil sample was collected at 3m depth from the ground surface in Chennur region at Kadapa district of Andhra Pradesh, India.

15.2.1.2 Bacteria

Alkaliphilic Bacteria isolates were collected from the Microbiology Department, Sri Venkateswara University, Tirupati. It is gram-positive bacteria received in the liquid form and aerobically re-energized in a sterilized high media nutrient broth overnight at 30°C.

Improvement of Properties of an Expansive Soil 121

15.3 Tests Conducted

15.3.1 Atterberg Limits

Atterberg Limits or Consistency Limits are important to classify the soil and understand the correlation between the limits and engineering properties of soils with respect to water content that include liquid limit, plastic limit, and shrinkage limit. The liquid limit and plastic limit values are determined by using Casagrande's apparatus according to the standard procedure laid in IS: 2720 (Part 5)–1985.

15.3.2 Unconfined Compression Strength Test

To analyze the shear strength parameters of the soil sample and soil-bacteria mixes, the Unconfined Compressive Strength test was performed at Optimum Moisture Content (OMC) and Maximum Dry Density (MDD) as per IS: 2720 (Part 10)–1991. The soil specimens of standard size (76mm height and 36mm diameter) were extracted and sheared at an axial strain rate of one percent (1%) per minute.

15.3.3 Soil Surface Morphology

$CaCO_3$ formation for bio-mineralization is analyzed using direct and microscopic observations using Scanning Electron Microscope studies. To conduct SEM analysis, expansive soil test samples are air-dried to remove moisture. Then, all the test samples are layered with gold and were analyzed in a Zeiss EV050 Scanning Electron Microscope (SEM) at 10kV.

15.3.4 pH Value

The acidic or alkaline characteristics of a soil sample can be quantitatively expressed by hydrogen ion activity commonly designated as pH. To conduct pH analysis, the soil sample is prepared in accordance with IS: 2720, (Part 1–(1983)).

15.4 Test Results

Expansive soils impose various structural problems due to their sensitiveness to slight changes in the moisture content. These soils pose problems on account of their high compressibility and low strength also. To reduce these problems, bacteria is induced to stabilize expansive soil. The Atterberg Limits and strength of the selected expansive soil is analyzed from the respective test results obtained from a series of tests on soil and soil-Alkaliphilic Bacteria mixtures at varying concentrations and curing times. The chemical and physical properties of tested soil samples are presented in Tables 15.1 and 15.2.

15.5 Plasticity Characteristics

Atterberg Limits are determined as behavior of soils is related to the amount of water. Atterberg Limits (liquid limit, plastic limit) of expansive soil are determined for soil alone and with the induction of Alkaliphilic Bacteria in different concentrations and at different curing time. Atterberg limits are found varying with the induction of varying concentrations of Alkaliphilic Bacteria into the natural soil. The liquid limit values of the expansive soil decreased with an increasing amount of bacteria concentration and a similar trend was found with the plasticity index as shown in Figures 15.1 and 15.3. The plastic limit values of expansive soil increased with an increasing amount of bacteria concentration and with curing times as shown in Figure 15.2.The plasticity index is a good indicator of swelling potential. The bio-cement aids flocculation and aggregates the clay particles that result in a decrease in Atterberg Limits of expansive soil. The results demonstrate, soil plasticity index and liquid limit experience a noticeable reduction after treating with bacterial concentration as depicted from Table 15.3.

TABLE 15.1

Properties of Tested Soil Sample

SI. No	Tested Soil	Properties
1.	Grain Size Distribution	
	(a) Gravel, (%)	3.6
	(b) Sand, (%)	10.4
	(c) Silt, (%)	40.0
	(d) Clay, (%)	46.0
2.	Atterberg Limits	
	(a) Liquid Limit, (%)	58.0
	(b) Plastic Limit, (%)	20.0
	(c) Plasticity Index, (%)	38.0
3.	Free Swell Index, (%)	110.0
4.	Differential Free Swell Index, (%)	10.0
5.	Swelling Pressure, kN/m^2	190.00
6.	Degree of Expansion	High
7.	Classification of Soil	CH (Clay with High Compressibility)
8.	Specific Gravity	2.65
9.	pH	8.33
10.	Compaction Characteristics	
	(a) Maximum Dry Density, (kN/m^3)	17.00
	(b) Optimum Moisture Content, (%)	15.5
11.	Unconfined Compressive Strength, (kPa)	233.00
12.	Shear Characteristics	
	(a) The angle of Internal friction(degrees)	16.0
	(b) Cohesion (kPa)	75.00
13.	Consolidometer or Oedometer Test	
	(a) Compression Index	0.346
14.	Coefficient of Permeability (cm/sec)	6.59×10^{-8}
15.	California Bearing Ratio	
	(a) CBR,(%) (unsoaked)	11.86
	(b) CBR,(%) (soaked)	6.27

TABLE 15.2

Chemical Properties of Tested Soil Samples

Sl. No	Tests Conducted	Chemical Properties
1.	Chlorides (ppm)	13.75
2.	Calcium Carbonate ($CaCO_3$) (%)	9.75
3.	Electrical Conductivity (dS/m)	0.1
4.	Potassium Oxide(K_2O) (kg/ha)	195.95
5.	Nitrogen (kg/ha)	50.18
6.	Organic Carbon (%)	0.14
7.	Organic matter	0.24
8.	Phosphates (P_2O_5) (kg/ha)	20.52
9.	Sulphates (%/mass)	82.31
10.	Sulphur (ppm)	54.00

Improvement of Properties of an Expansive Soil

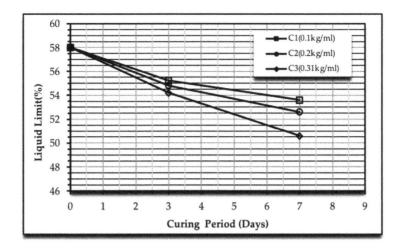

FIGURE 15.1 Variation of liquid limit values of soil with different concentrations of bacteria and curing time.

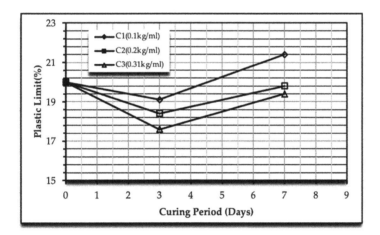

FIGURE 15.2 Variation of plastic limit values of soil with different concentrations of bacteria and curing time.

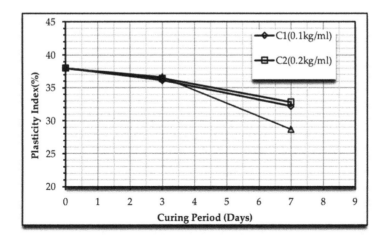

FIGURE 15.3 Variation of plasticity index values of soil with different concentrations of bacteria and curing time.

TABLE 15.3

Plasticity Characteristics of Expansive Soil Admixed with Different
Concentrations of Bacteria with Curing Time

Property	Liquid Limit (%)			Plastic Limit (%)			Plasticity Index (%)		
Concentration/Curing days	0	3	7	0	3	7	0	3	7
C_0(untreated soil)	58	58	58	20	20	20	38	38	38
C_1(0.1kg/ml)	55.2	53.6	48.4	19.1	21.4	22.9	36.1	32.2	25.5
C_2(0.2kg/ml)	54.8	52.6	50.2	18.4	19.8	22.1	36.4	32.8	28.1
C_3(0.31kg/ml)	54.2	50.6	49.2	17.6	19.4	21.9	36.6	28.7	27.3

TABLE 15.4

Unconfined Compressive Strength Values of Soil with Different
Concentrations Bacteria with Curing

Bacteria Concentration/Curing days	Unconfined Compressive Strength, (kPa)		
	0	3	7
C_0(untreated soil)	233.00	233.00	233.00
C_1(0.1kg/ml)	244.80	331.20	484.20
C_2(0.2kg/ml)	233.60	247.20	345.20
C_3(0.31kg/ml)	235.60	249.00	365.80

A graph of plastic limit vs curing period is plotted in y axis and x axis respectively for soils with different concentrations of bacteria named C1 (0.1kg/ml), C2 (0.2kg/ml), C3 (0.3kg/ml). Initially the liquid limit was same for all the soils i.e., 20%. At three days curing period, plastic limit of C1, C2, C3 was noted to be 19%, 18.5% and 17.8%. Which means all of them decreased? Then plastic limit started increasing with curing further. At 7 days the plastic limit of the soils C1, C2, C3 was observed to be 21.5%, 20% and 19.5%.

15.6 Strength

The variation of the strength of expansive soil specimens with different dosages of bacteria concentrations and different curing times (days) is shown in Figure 15.4. The Unconfined Compressive Strength of untreated expansive soil is 220kPa, and the Unconfined Compressive Strength values of treated expansive soil were 484.2kPa (0.1ml/kg), 345.2kPa (0.2ml/kg), and 365.8kPa (0.3ml/kg) for 7 days curing. Maximum Unconfined Compressive Strength value of 484.2kPa is obtained with 0.1ml/kg concentration for 7 days curing as shown in Figure 15.4. The Unconfined Compressive Strength values increased with increasing concentration of bacteria. This may be attributed to the biochemical reaction and the calcium which is present in the soil. It can be noted that the stiffness of the treated samples increased with treatment, and this could be due to the higher stiffness of the calcium precipitated.

15.7 pH Value

The pH values tests were conducted or soil-bacteria mix with curing for C_1 concentration only as the strength of selected soil is found to improve effectively with C_1 concentration of bacteria (0.1ml/kg). The pH value of the untreated soil is 8.27, which is moderately alkaline. The expansive soil treated sample

Improvement of Properties of an Expansive Soil

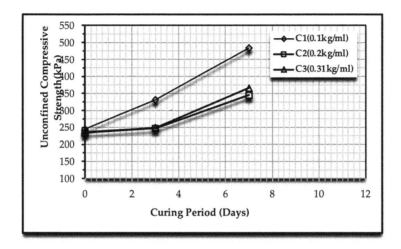

FIGURE 15.4 Variation of Unconfined compressive strength values of soil with different concentrations of bacteria and curing time.

TABLE 15.5
Effect of pH Values on UCS of Soil

No. of Days Curing	Bacteria Concentration (C_1)	
	pH	UCS(kPa)
0	8.46	244.8
3	8.62	331.2
7	9.12	484.2

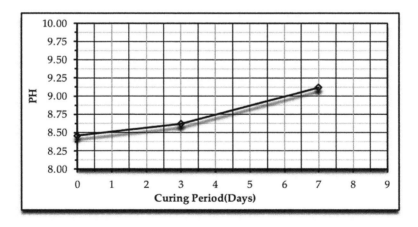

FIGURE 15.5 Variation of pH Values of soil mixed with C_1 concentration of bacteria on curing

pH values are 8.46, 8.62 and 9.12 respectively for 0 day, 3 day and 7 day curing. However, the pH value increased with 7 day curing as shown in Figure 15.5. The effect of pH values on Unconfined Compressive Strength values of treated soil is shown in Figure 15.5. The result of the study indicated that Alkaliphilic Bacteria can develop under the average temperature and can produce a high amount of calcite. As pH values increase, UCS values also increased as shown in Figure 15.6. pH values are effectively ion exchange resulting in binding the soil samples. The reduction in double-layer thickness (due to pH) leads to an extensive decrease in the soil–water system and improve soil strength.

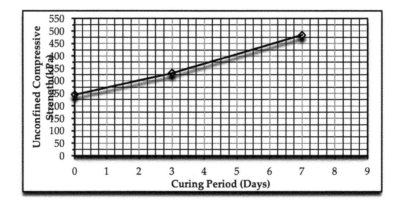

FIGURE 15.6 Variation of UCS values of soil mixed with C_1 concentration bacteria on curing

15.8 Micro Studies

Evaluation of morphology of the expansive soil with bacteria stabilizer was done using a Scanning Electron Microscope (SEM) in the form of micrographs as shown in Figure 15.7. The SEM micrographs depict the bio-cement-based mechanism in the stabilization of the expansive soil. Expansive soil is a very composite system where it is not easy to observe directly and can be easily observed using SEM. However, soil holes surrounded with rod-shaped bacteria cells are seen with crystal structure with curing for 0, 3 and 7 days for the soil-bacteria mix samples. Figure 15.7a, 15.7b, 15.7c and 15.7d show micrographs from the SEM for treated soils at magnifications of 4.00 KX. Figure 15.7 a represents expansive soil without mixing bacteria. Figure 15.7b, 15.7c and 15.7d show white surfaces indicating the $CaCO_3$

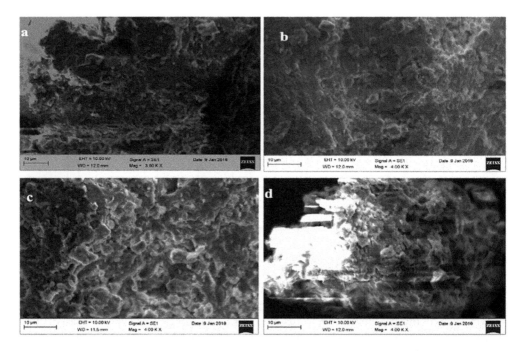

FIGURE 15.7 SEM micrograph (a) Soil sample, (b) Soil with induced bacteria (C_1 concentration) at 0 day curing, (c) Soil with induced bacteria (C_1 concentration) after 3 days curing, (d) Soil with induced bacteria (C_1 concentration) after 7 days curing.

precipitation with increasing curing times. The results were in agreement with Cheng et al. (2013) and Sham et al. (2013) where calcite-precipitating bacteria lead to the formation of different forms of calcium carbonate that enhances the strength of the soil. Natural expansive soil samples showed typical soil morphology (Figure 15.7a), in concurrence with Goodarzi et al. (2016). The microstructural change with the addition of bacteria and with curing time for treated samples promotes the bonding between particles which contributes to the strength development and reduction in pore volume. During the curing period, the volume of pores becomes smaller due to the hydration reaction. Figure 15.7b, 15.7c and 15.7d show the presence of different clusters affected by bio-cloggation reaction. These parts have different micronics sizes due to their reactions. The SEM analysis revealed the elemental composition in the soil after stabilization and curing period. Before treatment, the porosity of the samples is high. However, a denser structure is evident after treatment.

15.9 Conclusions

The effect of Alkaliphilic Bacteria induction into the selected expansive soil (CH-High Compressible Clay) with curing times on plasticity characteristics and Unconfined Compressive Strength is tested in this paper. On the basis of the test results, induction of bacterial concentration resulted in reduction of liquid limit and plasticity index of expansive soil and increment in the Unconfined Compressive Strength. The decrement in plasticity index was very high. The average decrease in plasticity index is 34.6% and the percentage increase in Unconfined Compressive Strength is 90%. Bacteria induction has a higher cementing effect in soil due to decrease in percentage of clay 80% to 45%. The pH value of soil-bacteria mixtures increased, from moderately alkaline (pH = 8.27) to highly alkaline range (pH = 9.12), helping in increasing the bacteria colonies in the soil, thereby increasing the soil strength. Microstudies using (SEM) analysis depict closing of pores of soil with $CaCO_3$ content on curing and improving soil strength. The use of bacteria for the improvement of expansive soil characteristics is not only cost-effective method but also an eco-friendly and simple method.

REFERENCES

Basha, S., Lingamgunta, L.K., Kannali, J., Gajula, S.K., Bandikari, R., Dasari, S., & Balaji, V.K. (2018). Subsurface endospore-forming bacteria possess bio-sealant properties. *Scientific Reports*, 8(1), 6448.

Benini, S., Rypniewski, W.R., Wilson, K.S., Miletti, S., Ciurli, S., & Mangani, S. (1999). A new proposal for urease mechanism based on the crystal structures of the native and inhibited enzyme from Bacillus pasteurii: Why urea hydrolysis costs two nickels. *Structure*, 7(2), 205–216.

Burbank, M., Weaver, T., Lewis, R., Williams, T., Williams, B., & Crawford, R. (2013). Geotechnical tests of sands following bio-induced calcite precipitation catalyzed by indigenous bacteria. *Journal of Geotechnical and Geoenvironmental Engineering*, 139(6), 928–936.

Burbank, M. B., Weaver, T. J., Green, T. L., Williams, B. C., & Crawford, R. L. (2011). Precipitation of calcite by indigenous microorganisms to strengthen liquefiable soils. *Geomicrobiology Journal*, 28(4), 301–312.

Chittoori, B. C., Burbank, M., & Islam, M. T. (2018). *Evaluating the effectiveness of soil-native bacteria in precipitating calcite to stabilize expansive soils. International Foundation Congress and Equipment Expo 2018*: March 5–10, 2018 Orlando, Florida, published by the American Society of Civil Engineers (ASCE). doi: 10.1061/9780784481592.007.

Chittoori, B., & Neupane, S. (2018). *Evaluating the Application of Microbial Induced Calcite Precipitation Technique to Stabilize Expansive Soils*. GeoChina, Hang Zhou, China.

Cheng, L., & Shahin, M. A. (2016). Urease active bioslurry: A novel soil improvement approach based on microbially induced carbonate precipitation. *Canadian Geotechnical Journal*, 53(9), 1376–1385.

Chen, F. H. (1988). *Foundations on Expansive Soils*. Elsevier, New York.

Cheng, L. & Cord-Ruwisch, R. (2012). In situ soil cementation with ureolytic bacteria by surface percolation. *Ecological Engineering* 42, 64–72

Cheng, L., Cord-Ruwisch, R. & Shahin, M.A. (2013). Cementation of sand soil by microbially induced calcite precipitation at various degrees of saturation. *Canadian Geotechnical Journal* 50, 81–90.

Chu, J., Stabnikov, V., & Ivanov, V. (2012). Microbially induced calcium carbonate precipitation on surface or in the bulk of soil. *Geomicrobiology Journal*, 29(6), 544–549.

DeJong, J.T., Soga, K., Banwart, S.A., Whalley, W.R., Ginn, T.R., Nelson, D.C., & Barkouki, T. (2010). Soil engineering in vivo: harnessing natural biogeochemical systems for sustainable, multi-functional engineering solutions. *Journal of the Royal Society Interface*, 8(54), 1–15.

Douglas, S., & Beveridge, T.J. (1998). Mineral formation by bacteria in natural microbial communities. *FEMS Microbiology Ecology*, 26(2), 79–88.

Goodarzi, A.R., Akbari, H.R., & Salimi, M. (2016). Enhanced stabilization of highly expansive clays by mixing cement and silica fume. *Applied Clay Science* 132–133, 675–684.

Hamdan, N., Kavazanjian, E. Jr., Rittmann, B.E., & Karatas, I. (2011). Carbonated mineral precipitation for soil improvement through microbial denitrification. In: *Proceedings of the GeoFrontiers 2011: AdvancesinGeotechnicalEngineering*, Dallas, TX, ASCE Geotechnical Special Publication 211, 3925–3934.

Hammes, F., & Verstraete, W. (2002). Key roles of pH and calcium metabolism in microbial carbonate precipitation. *Reviews in Environmental Science and Biotechnology*, 1(1), 3–7.

Ivanov, V., & Chu, J. (2008). Applications of microorganisms to geotechnical engineering for bioclogging and biocementation of soil in situ. *Reviews in Environmental Science and Bio/Technology*, 7(2), 139–153.

Li, M., Li, L., Ogbonnaya, U., Wen, K., Tian, A., & Amini, F. (2016). Influence of fiber addition on mechanical properties of MICP-treated sand. *Journal of Materials in Civil Engineering*, 28(4), 04015166.

Lloyd, A.B., & Sheaffe, M.J. (1973). Urease activity in soils. *Plant and Soil*, 39(1), 71–80.

Montoya, B.M., & DeJong, J.T. (2015). Stress-strain behavior of sands cemented by microbially induced calcite precipitation. *Journal of Geotechnical and Geoenvironmental Engineering*, 141(6), 04015019.

Martinez, B.C., DeJong, J.T., Ginn, T.R., Mortensen, B.M., Barkouki, T.H., Hunt, C., Tanyu, B., & Major, D. (2013). Experimentaloptimizationofmicrobialinducedcarbonateprecipitationforsoilimprovement. *ASCEJ Journal of Geotechnical &Geoenvironmental Engineering* 139(4), 587–598.

Mortensen, B.M., Haber, M.J., DeJong, J.T., Caslake, L.F., & Nelson, D.C. (2011). Effects of environmental factors on microbial induced calcium carbonate precipitation. *Journal of Applied Microbiology*, 111(2), 338–349.

Obuzor, G.N., Kinuthia, J.M., & Robinson, R.B. (2011). Enhancing the durability of flooded low-capacity soils by utilizing lime-activated ground granulated blastfurnace slag (GGBS). *Engineering Geology*, 123(3), 179–186.

Thyagaraj, T & Pusadkar, S.S. (2019). Soil improvement using microbial: A review. In *Ground Improvement Techniques and Geosynthetics* (329–335). Springer, Singapore.

Sham, E., Mantle, M.D., Mitchell, J., Tobler, D.J., Phoenix, V.R., & Johns, M.L. (2013). Monitoring bacterially induced calcite precipitation in porous media using magnetic resonance imaging and flow measurements. *Journal of Contaminant Hydrology*, 152, 35–43.

Van der Star, W.R.L., VanWijngaarden, W.K., Van Paassen, L.A., VanBaalen, L.R., & Zwieten, G. (2011). *Stabilization of gravel deposits using microorganisms*. In *Proceedings of the 15th European Conference on Soil Mechanics and Geotechnical Engineering*, Athens, Greece, IOS Press, 5–9 October 2011.

VanWijngaarden, W.K., Vermolen, F.J., VanMeurs, G.A.M., & Vuik, C. (2011). Modelling biogrout: a new ground improvement method based on microbial-induced carbonate precipitation. *Transport in Porous Media*, 87(2), 397–420.

16

Application of Treated Mixed Fruit Wastes in Soil Stabilization

David O. Olukanni and Chukwume G. Ijeh
Covenant University, Ota, Nigeria

CONTENTS

16.1 Introduction 129
16.2 Methodology 131
16.3 Atterberg Limits 133
16.4 Conclusion.. 135
References.. 135

16.1 Introduction

Management of waste has become an issue of concern probing all nations across the globe on its effective management (Amasuomo and Baird, 2016). In developing countries, a small fraction of these wastes is being collected and managed. Governments of nations are constantly seeking better means of processing and disposal of these wastes (Schiopu et al., 2007; Olukanni et al., 2016; Olukanni & Aremu, 2017). Due to its large population of over 180 million people, Nigeria has been described to be one of the largest producers of solid waste in Africa (Aneke and Attah, 2016; Bakare, 2018). In addition, wastes generated in Nigeria have been estimated to be about 32 million tonnes and only about 20–30% is collected by appropriate waste collection authorities, which, consequently, has made open-dumping the next option adopted by the masses (Olukanni et al., 2016; Olukanni et al., 2017; Bakare, 2018). In line with this, solid waste management has developed into one of the most crucial and pressing environmental challenges faced by both urban and rural areas of Nigeria.

Open-dumping poses a major threat to a number of environmental processes. Leachate from open-dumping could drastically change the quality of water in the groundwater table (Olukanni et al., 2016; Olukanni et al., 2017). In Nigeria, boreholes are largely being depended on for the supply of water, which require little or no treatment. However, with constant pollution of groundwater from dumpsites, there would be greater need to treat water, which, in turn, means greater cost for the economy. Heaps and piles of organic wastes also generate gases such as carbon dioxide and methane gas which are both greenhouse gases (Olukanni & Aremu, 2017). These gases have been revealed to contribute to the greenhouse effect and in turn, climate change.

Environmental pollution and greenhouse gas (GHG) emissions are the results of indiscriminate disposal, dumping and inadequate methods of processing these wastes (Mia et al., 2018; Olukanni and Oresanya, 2018). Wastes are generally regarded as "useless" and unwanted material(s) which are usually discarded by some means. However, certain wastes can become "useful" through resource recovery if they are removed from the waste stream and made to undergo certain processes (Oyelola and Babatunde, 2008). Managing waste has therefore been considered a key component in improving public health and environmental protection (Götze et al., 2016). Although the European Waste Framework Directive has quoted prevention as the best management practice that could possibly be adopted, there is still an

DOI: 10.1201/9781003217619-18

129

extensive way to go with wastes that have already been generated, especially in developing countries (Götze et al., 2016).

The critical components to be considered therefore, is the processing and disposal of wastes that have been generated. Different researchers such as (Alvarenga et al., 2015; Gutiérrez et al., 2017, Olukanni & Aremu, 2017; Olukanni et al., 2017; Olukanni & Nwafor, 2019), have all been advocates for composting as a method of managing organic wastes. Composting is aimed at improving the soil properties for agricultural use. Here, organic waste can be aerobically or anaerobically digested and used on agricultural farmland as fertilizer for crops. This process produces a dark-brown humus-like substance. Another waste management option is biogas which is produced from the anaerobic decomposition of organic waste and is particularly useful in the management of sewage sludge. This is because the treatment of sewage sludge is very expensive and its decomposition produces a high percentage of gas (Verdaguer et al., 2016). Biogas serves as a good substitute for cooking gas because it is cheap and can be domestically prepared.

The proper management of waste is vital and the useful resource in organic waste should not just be left fallow when it can be put to use. With this in mind, it is common to come across rather unsatisfying soils in the course of road construction, hence the need to stabilize the soil. The use of some chemical additives such as bitumen emulsion, cement and lime has been previously, and is currently being used as soil stabilizers. However, some of these stabilizers such as the bitumen emulsion are not considered sustainable. As a result of runoff, some of its particles may find its way into nearby water bodies causing major pollution to the aquatic life and a setback to all who rely on the source of water for one purpose or the other. This poses a major threat to the natural ecosystem. In addition, cement and lime are very expensive stabilizers. Production of cement causes extreme air pollution and is a health threat to all who might live close to such areas. With the world's paradigm shift, "greener" and more sustainable ways is constantly being sort after in order to leave more resources for the upcoming generation.

The likelihood of minimizing the use of degenerating resources in construction, together with the necessity for the suitable disposal of organic waste in the environment, makes reuse of organic materials attractive in economic, engineering and environmental terms (Satoshi et al., 2010). Organic waste has not been utilized to its maximum potential, especially in developing countries (Olukanni and Aremu, 2017). Organic wastes are being indiscriminately disposed of and need a more effective and efficient ways of management. These organic wastes have so much resource which could make them beneficial to soils, construction and the society as a whole. Alternative materials for soil stabilization have been tested. These materials include palm bunch, almond shells, groundnut shell ash (GSA), fly ash (FA); wood ash (WA); sawdust (SD); bagasse fiber (BF); rise husk ash (RHA); palm fibers (PF); municipal solid waste (MSW), high density polyethylene (HDPE) and glass (Phanikumar & Sharma, 2004; Okagbue, 2007; Marandi and Bagheripour, 2008; Prakash & Sridharan, 2009; Folagbade and George, 2010; Canakci et al., 2015; Butt et al., 2016; Butt et al., 2016; Dang et al., 2016; Campuzano & González-Martínez, 2016; Wafa et al., 2018). Some of these materials have been considered acceptable for soil stabilization and these studies have helped broaden the knowledge on the effectiveness of different waste materials as soil stabilizers. However, the study of using specifically fruit waste as a soil stabilizer has been limited.

It has been reported that there is increasing cost in the use of conventional stabilizers, such as cement and lime, to improve the quality of soil for engineering purposes. Soil stabilization is the alteration of soils either through a physical process or the addition of an additive to enhance the soil's physical properties (Wafa et al., 2018). This option seeks to improve the properties of the soil for engineering purposes. Soil stabilization can also be described as a procedure in which existing properties of soil are altered in order to meet an engineering need (Kalsekar et al., 2016). It is used mainly when weak soils are encountered in the course of construction such as low bearing capacity, high shrink-swell tendencies and the likes.

The majority of waste being produced in Covenant University is organic waste and it constitutes about 45% of the total generated waste (Okeniyi & Udonwan, 2016; Olukanni et al., 2018). The main aim of this research is to find an alternative, effective and sustainable means of managing organic wastes, specifically fruit wastes, in soil stabilization. More specifically, is to assess some organic wastes being generated in the institution, carry out experimental study on the use of the waste as a soil stabilizer and determine the level of effectiveness of the waste as a soil stabilizer.

16.2 Methodology

Fruit Wastes (FW) was collected from Cafeterias 1 and 2 of Covenant University. The FW was sorted to remove any unwanted material such as nylon bags. FW such as banana peel (BP), orange peel (OP), pineapple peel (PP) and watermelon rind (WR) were used (Figure 16.1a). Disturbed soil sample was also collected from Covenant University on latitude and longitude of 6° 40' 17" N / 3° 9' 29" E, respectively. The choice of location was an eroded area beside the College of Science and Technology (CST) of the university where all drainage water in the university is directed.

After sorting, the FW was oven-dried for a period of 72hrs in the Geotechnical Laboratory of the Civil Engineering Department, Covenant University at a temperature of 100°C. Figure 16.1b shows a portion of the oven-dried fruit wastes. After drying, the FW were ground into smaller sizes to form FWP in order to aid mixing with the soil sample (Figure 16.1c).

Different percentages of fruit waste (Fruit Waste Powder, FWP) were added to the soil sample at 5%, 10% and 15% of soil. Atterberg's Limit Test, Compaction Test, California Bearing Ratio (CBR) and Unconfined Compressive Strength (UCS) were all determined using ASTM specifications. This also includes the determination of the Optimum Moisture Content (OMC), and the Maximum Dry Density (MDD). The soil samples were prepared using the mixture percentages in Table 16.1.

The procedures were carried out for FWP at 0%, 5%, 10% and 15% of soil sample. The test procedures for each test mentioned above were carried out according to standard procedures as presented in Table 16.2.

FIGURE 16.1 (a) and (b): Fruit Wastes (FW); (c) Fruit Waste Powder (FWP)

TABLE 16.1

Mixture Percentages of Soil and FWP

S/N	% of Soil Sample	% of FWP
1.	100%	0%
2.	95%	5%
3.	90%	10%
4	85%	15%

TABLE 16.2

Laboratory Tests and Standards

Test	Specification
Particle Size Distribution	ASTM D6913/D6913M
Atterberg's Limit Test	ASTM D4318
Compaction Test	ASTM D698
California Bearing Ratio (CBR)	ASTM D1883-05
Unconfined Compressive Strength (UCS)	AS 5101.4-2008

In order to find the liquid limit (LL) of the sample, about 150–200g of sample passing 425µm IS (International Sieves) sieve was taken. The soil sample was then mixed thoroughly with water till it was moist, in order to get about 25–35 blows for first trial. About 20g of this same sample was kept aside for plastic limit (PL) testing. For the second trial, more water was added, so as to get about 20–30 blows. Lastly, more water was also added to the third trial, sufficient to get about 15–25 blows. For each trial, soil sample was taken for moisture content determination (Mishra, 2018a, 2018b, 2018c, 2018d).

For the Proctor compaction test, 20kg of soil passing through 20mm and 4.7mm IS sieves was mixed thoroughly with water. About 2.5kg of the soil mixed with water was then added to the compaction mold in three (3) layers. Each layer was given 25 uniformly distributed blows. Samples from the top, middle and bottom of the compaction mold were taken in order to find the moisture content. The experiment was repeated three (3) times and the dry density for each trial was calculated.

Determination of the CBR required 5kg of soil sample passing 20mm IS sieves to be mixed thoroughly with the OMC of the soil. The sample was then added to the mold in five (5) layers with each layer getting 56 evenly distributed blows. The mold was mounted on the testing machine and the load corresponding to various penetrations were noted. The CBR at 2.5mm and 5.0mm (using Equation 16.1 and Table 16.2) were calculated and the higher value was chosen.

$$California\ Bearing\ Ratio\left(CBR\right)=\frac{Test\ Load}{Standard\ Load} \tag{16.1}$$

To calculate the UCS of each sample, soil sample was collected and saturated using a sampling tube. The sample was then transferred to a split mold; the split mold was then removed after carefully placing the sample. The length and diameter of the sample was then measured and placed in the compression machine. Compression load was applied on the sample and the dial readings were recorded for various percentages of strain until failure occurred. Samples were then taken around the failure zone for moisture content determination.

Lateritic soil with different percentage of Fruit Waste Powder (FWP) was used for soil stabilization. Here, discussion on test results of Proctor Compaction-, California Bearing Ratio (CBR), Atterberg's Limit- and Unconfined Compressive Strength-test on soil with 0%, 5%, 10% and 15% of FWP through test reading are presented in Table 16.3. As FWP is varying in percentage, properties of soil are also changing. Proctor Compaction test was carried out on soil with various doses of FWP to determine the

Application of Treated Mixed Fruit Wastes

TABLE 16.3

Standard Loads for Penetrations of 2.5mm and 5.0mm

Penetration (mm)	Standard Load (kg)
2.5	1,370
5.0	2,055

Optimum Moisture Content (OMC) and Maximum Dry Density. This test was carried out to check their moisture-density relationships. Table 16.3 shows the variation in the OMC and MDD with varied percentages of FWP.

From Table 16.3, it can be observed that the addition of FWP to the soil caused a decrease in the MDD. This can be attributed to the relatively low density of FWP when compared to the soil sample. Also, a study done by Dutta & Rao (2007) showed that the presence of an additive in the soil makes the soil offer a higher resistance to compaction, thereby resulting in low density as the quantity of the additive increases. A study done by Butt et al. (2016) also showed a reduction in the MDD with increase in sawdust ash (SDA) from 0% to 12%. It can also be seen from Table 16.3 that the OMC increased with increase in the FWP in the soil sample. This is due to the absorbent nature of the FWP. This means that the FWP has a high affinity for water and does not repel it. Butt et al. (2016) also showed an increase in OMC with the addition of SDA. This therefore means that, the higher the amount of FWP in the soil, the more expansive the soil becomes due to the absorption of more water.

16.3 Atterberg Limits

Atterberg limits or consistency limits are used to explain the expansivity or the plastic nature of soil. This test was carried out to demonstrate the effect of FWP on the consistency of the soil. The variations of the PL, LL, and PI with the FWP content are presented in Table 16.4. It was observed that the addition of FWP at 10% increased the LL of the soil by 10.1% and a subsequent reduction of 1.1% at 15% addition. With increase in the LL of the soil, there was also an increase in its compressibility and shrink-swell potential. For the PL, increase in the PL indicates increase in the plasticity of the soil. The plasticity of the natural soil increased at 10% from 6.9% to 19.4%. The PL subsequently reduced to 19.0% due to more addition of FWP. The PI is highest at 10% due to high PL and LL values. A study done (Butt et al., 2016), showed similar effects of SDA on soil. Because clay particles carry a negative charge and equally attract the positive charge on water. Therefore, a reduction in the LL of the soil was as a result of a reduction in the clay content of the soil and vice versa. It can then be said that the clay content of the soil reduced at 15% addition of FWP.

The addition of FWP to the soil caused a decrease in the MDD. This can be attributed to the relatively low density of FWP when compared to the soil sample. The presence of an additive in the soil makes the soil offer a higher resistance to compaction, thereby resulting in low density as the quantity of the additive increases. A study done by Butt et al. (2016) also showed a reduction in the MDD with increase in

TABLE 16.4

MDD and OMC of Soil Sample at Different Percentages of FWP

% of FWP	Maximum Dry Density (MDD), kg/m³	Optimum Moisture Content (OMC), %
0%	1,913.0	15.4
5%	1,737.5	20.1
10%	1,661.4	22.0
15%	1,554.5	24.0

sawdust ash (SDA) from 0% to 12%. It was also observed from Table 16.4 that the OMC increased with increase in the FWP in the soil sample. This is due to the absorbent nature of the FWP. This means that the FWP has a high affinity for water and does not repel it. Butt et al. (2016) also showed an increase in OMC with the addition of SDA. This therefore means that, the higher the amount of FWP in the soil, the more expansive the soil becomes due to the absorption of more water.

It was observed that the addition of FWP increased the LL of the soil by 10.1% at 10% and a subsequent reduction of 1.1% at 15%. With increase in the LL of the soil, there is also an increase in its compressibility and shrink-swell potential. The plasticity of the natural soil increased at 10% from 6.9% to 19.4%. The increase in the PL indicates increase in the plasticity of the soil. The PL subsequently reduced to 19.0% due to more addition of FWP. The PI is highest at 10% due to high PL and LL values. A study done by Butt et al. (2016) showed similar effects of SDA on soil. Therefore, a reduction in the LL of the soil is as a result of a reduction in the clay content of the soil and vice versa. It can then be said that the clay content of the soil at 15% reduced. Table 16.5 shows the CBR values at different percentages of FWP addition.)

Generally, from the results obtained, it can be said that with an increase in the percentage of FWP, both the liquid limit (LL) and the plastic limit (PL) increased till they both reached a maximum value at 10% and finally decreased at 15%. It can also de deduced that the maximum plasticity index (PI) occurred at

TABLE 16.5

Result Summary of the Tests Performed on the Samples

Test	Natural Soil 0% FWP	5% FWP	10% FWP	15% FWP
Optimum Moisture Content (OMC), %	15.4	20.1	22.0	24.0
Maximum Dry Density (MDD), kg/m^3	1,913.0	1,737.5	1,661.4	1,554.5
Liquid Limit (LL), %	22.0	27.8	32.1	31.0
Plastic Limit (PL), %	12.5	17.9	19.4	19.0
Plasticity Index (PI), %	9.5	9.9	12.7	12.0
California Bearing Ratio (CBR), %	2.35	2.69	3.55	2.86
Unconfined Compressive Strength (UCS), kN/m^2	5.05	14.00	17.00	6.10

TABLE 16.6

CBR Values at Different Percentages of FWP

Penetration, mm = Readings × 0.01mm	Load, kg = Proving Ring Reading × 1.176kg				CBR, %			
	0%	5%	10%	15%	0%	5%	10%	15%
0.5	5.880	7.056	11.760	9.408				
1.0	9.408	11.760	17.640	12.936				
1.5	16.464	17.640	25.872	19.992				
2.0	22.344	25.872	35.280	25.872				
2.5	28.224	31.752	47.040	34.104	2.06	2.32	3.43	2.49
3.0	35.280	37.632	51.744	37.632				
3.5	42.336	43.512	57.624	44.688				
4.0	45.864	48.216	62.328	49.392				
5.0	48.216	55.272	72.912	58.800	2.35	2.69	3.55	2.86
6.0	51.744	61.152	82.320	68.208				
7.0	55.272	64.680	91.728	74.088				
8.0	58.800	70.560	97.608	81.144				
9.0	61.152	74.088	101.136	87.024				
10.0	63.504	76.440	105.840	94.080				

Application of Treated Mixed Fruit Wastes 135

TABLE 16.7

UCS of Soil Sample at Different Percentages of FWP

S/N	Percentage proportion of stabilization	Optimum Moisture Content (%)	Bulk density, Y_b (mg/m³)	Maximum Dry Density, Y_d (mg/m³)	Unconfined Compressive Strength, qu (kN/m²)	Cohesion, qu/2 (kN/m²)
1	100% soil + 0% FWP	15.0	2.185	1.900	5.05	2.03
2	95% soil + 5% FWP	20.0	2.076	1.730	14.0	7.0
3	90% soil + 10% FWP	22.0	2.025	1.660	17.0	8.5
4	85% soil + 15% FWP	24.0	1.925	1.552	6.1	3.1

10%. It was observed that the CBR at 5% FWP increased by a value of 0.34% to give a value of 2.69%. The CBR also increased at 10% FWP to give 3.55%. This value was taken as optimum because the CBR at 15% then decreased although not as low as the CBR of the natural soil which corroborates the result from Butt et al. (2016) with SDA at 8%. Comparing the CBR values to the LL, PL and PI, it can be said that the highest CBR value was achieved at the highest of the LL, PL and PI.)

Table 16.7 shows the results for UCS of the soil samples at varying percentages. The UCS is basically used to check the performance of the additive on the bearing capacity of the soil. The higher the UCS of the soil, the higher the bearing capacity of the soil. UCS is used to determine the bearing capacity of soils for any type of construction. It was observed that the lowest UCS value was gotten for the natural soil. The UCS increased by 1.5kN/m² to give 7.0kN/m² and 8.5kN/m², respectively. The UCS at 15% then dropped to a value of 3.1kN/m² which could be as a result of the drop in the clay content.

16.4 Conclusion

In conclusion, it was found that an increase in the percentage of FWP caused an increase in the Optimum Moisture Content (OMC), liquid limit (LL), plastic limit (PL) and a decrease in the Maximum Dry Density (MDD). Due to increase in the OMC of the soil at different percentages of FWP, it can be said that the water absorption and expansive nature of the soil also increased. LL and PL reached an optimum value at 10% and decreased at 15%. The plasticity index also increased and reached its optimum at 10% with a value of 12.4%.

The California Bearing Ratio (CBR) and the Unconfined Compressive Strength (UCS) increased by a relatively sufficient amount to a value of 3.55% and 8.5kN/m² at 10%, respectively. However, the CBR and UCS of the soil at the various percentages did not experience a very generous increase. It can therefore be concluded that FWP can only be used as a soil stabilizer for areas with low wheel load or low volume traffic for subgrade in the ratio of 1:9 (FWP: soil). This study is therefore important in adding to the existing knowledge. In addition, management of fruit wastes would be more effective, and its disposal would reduce tremendously. Construction costs would also reduce and more importantly, the environment would be kept safe.

REFERENCES

Agriculture Victoria. (2017). What is soil. Retrieved April 26, 2019, from http://agriculture.vic.gov.au/agriculture/farm-management/soil-and-water/soils/what-is-soil.

Alvarenga, P., Mourinha, C., Farto, M., Santos, T., Palma, P., Sengo, J., and Cunha-Queda, C. (2015). Sewage sludge, compost and other representative organic wastes as agricultural soil amendments: Benefits versus limiting factors. *Waste Management*, 40(276): 44–52.

Amasuomo, E., and Baird, J. (2016). The concept of waste and waste management. *Journal of Management and Sustainability*, 6(4): 88.

Aneke, R. I., and Attah, E. Y. (2016). Waste management and sustainable development in Nigeria: A study of Anambra state waste management agency. *European Journal of Business and Management,* 8(17): 132–144.

Bakare, W. (2018). *Solid Waste Management in Nigeria*. BioEnergy Consult. Retrieved January 20, 2019, from: https://www.bioenergyconsult.com/tag/solid-waste-management-in-nigeria/.

Butt, W. A., Gupta, K., and Jha, J. N. (2016). Strength behavior of clayey soil stabilized with saw dust ash. *International Journal of Geo-Engineering*, 7(18): 1–9.

Campuzano, R., and González-Martínez, S. (2016). Characteristics of the organic fraction of municipal solid waste and methane production: A review. *Waste Management*, 54(1): 3–12.

Canakci, H., and Aziz, A., Celik, F. (2015). Soil stabilization of clay with lignin, rice husk powder and ash. *Geomechanics and Engineering*, 8(1): 67–79.

Dang, L. C., Fatahi, B., and Khabbaz, H. (2016). Behaviour of expansive soils stabilized with hydrated lime and bagasse fibres. Advances in Transportation Geotechnics 3, The 3rd International Conference on Transportation Geotechnics (ICTG), Portugal. 143(1): 658–665.

Folagbade, O., and George, M. (2010). Groundnut shell ash stabilization of black cotton soil. *Electronic Journal of Geotechnical Engineering*, 15(1): 415–428.

Götze, R., Astrup, T. F., Scheutz, C., and Boldrin, A. (2016). Composition of waste materials and recyclables. Technical University of Denmark, DTU Environment, Denmark.

Gutiérrez, M. C., Siles, J. A., Diz, J., Chica, A. F., and Martín, M. A. (2017). Modelling of composting process of different organic waste at pilot scale: Biodegradability and odor emissions. *Waste Management*, 59(1): 48–58.

Marandi, S. M., and Bagheripour, M. H. (2008). Strength and ductility of randomly distributed palm fibres. *American Journal of Applied Sciences*, 5(3): 209–220.

Mia, S., Uddin, M. E., Kader, M. A., Ahsan, A., Mannan, M. A., Hossain, M. M., and Solaiman, Z. M. (2018). Pyrolysis and co-composting of municipal organic waste in Bangladesh: A quantitative estimate of recyclable nutrients, greenhouse gas emissions, and economic benefits. *Waste Management*, 75(1): 503–513.

Mishra, G. (2018a). *California bearing ratio test on subgrade soil.* The Constructor - Civil Engineering Home. Retrieved April 13, 2019, from: https://theconstructor.org/geotechnical/california-bearing-ratio-test/2578/.

Mishra, G. (2018b). Determination of plastic limit of soil. The Constructor - Civil Engineering Home. Retrieved April 7, 2019, from: https://theconstructor.org/geotechnical/determination-plastic-limit-soil/2929/.

Mishra, G. (2018c). Determining unconfined compressive strength of cohesive soil. The Constructor - Civil Engineering Home. Retrieved April 14, 2019, from: https://theconstructor.org/geotechnical/unconfined-compressive-strength-of-cohesive-soil/3134/.

Mishra, G. (2018d). Proctor soil compaction test. The Constructor - Civil Engineering Home. Retrieved April 3, 2019, from: https://theconstructor.org/geotechnical/compaction-test-soil-proctors-test/3152/.

Okagbue, C. O. (2007). Stabilization of clay using woodash. *Journal of materials in Civil Engineering, ASCE*, 19(1): 14–18.

Olukanni, D. O. Adeleke, J. O., and Aremu, D. D. (2016). A review of local factors affecting solid waste collection in Nigeria. *Pollution*, 2(3): 339–356.

Olukanni, D. O. and Aremu, D. (2017). Provisional evaluation of composting as priority option for sustainable waste management in South-West Nigeria. *Pollution*, 3(3): 417–428.

Olukanni, D. O., Olujide, J. A., and Kehinde, E. O. (2017). Evaluation of the impact of dumpsite leachate on groundwater quality in a residential institution in Ota, Nigeria. *Covenant Journal of Engineering & Technology (CJET)*, 1(1): 18–33.

Olukanni, D. O., and Oresanya, O. O. (2018). Progression in waste management processes in Lagos state, Nigeria. *Journal of Engineering Research in Africa (JERA)*, 35: 11–23.

Olukanni, D. O., Aipoh, A. O., and Kalabp, I. H. (2018). Recycling and reuse technology: Waste to wealth initiative in a private tertiary institution, Nigeria. *Recycling*, 3(44): 1–12.

Olukanni, D. O., and Nwafor, C. (2019). Public-private sector involvement in providing efficient solid waste management services in Nigeria. *Recycling*, 4(19): 1–9.

Oyelola, O., and Babatunde, A. (2008). Characterization of domestic and market solid wastes at source in Lagos metropolis, Lagos, Nigeria. *African Journal of Environmental Science and Technology*, 3(12): 430–437. doi:10.1016/j.jpeds.2012.06.042.

Phanikumar, B. R., and Sharma, R. S. (2004). Effect of fly ash on engineering properties of expansive soils. *Journal of Geotechnical and Geo-environmental Engineering, ASCE*, 130(7): 764–767.

Prakash, K., and Sridharan, A. (2009). Beneficial properties of coal ashes and effective solid waste management, practice periodical of hazardous, toxic, and radioactive waste management. *African Society Civil Engineers*, 13(4): 239–248.

Schiopu, A. M., Apostol, I., Hodoreanu, M., and Gavrilescu, M. (2007). Solid waste in romania: Management, treatment and pollution prevention practices. *Environmental Engineering and Management Journal*, 6(5): 451–465.

Verdaguer, M., Molinos-Senante, M., and Poch, M. (2016). Optimal management of substrates in anaerobic co-digestion: An ant colony algorithm approach. *Waste Management*, 50(1): 49–54.

Wafa, B., Bashir, A., Nabi, S. Z., and Illahi, U. (2018). An experimental investigation of soil stabilized with almond shells: A tenable solution. *International Journal of Advance Research in Science and Engineering*, 7(4): 528–543.

17

Development of Flexible Pavement Cost Models for Weak Subgrade Stabilized with Fly Ash and Lime

Satya Ranjan Samal and Malaya Mohanty
KIIT Deemed to be University, Bhubaneswar, India

CONTENTS

17.1 Introduction 139
17.2 Methodology 140
    17.2.1 Fly Ash 140
    17.2.2 Use of Lime 140
    17.2.3 SPSS and Cost Modeling 140
17.3 Specifications of IRC 141
    17.3.1 Subgrade Soil 141
    17.3.2 Liquid Limit 141
    17.3.3 Plasticity Index 141
    17.3.4 Density Requirement 141
    17.3.5 CBR 141
17.4 Economic Analysis 143
17.5 Discussion and Conclusion 143
References 144

17.1 Introduction

Construction methods should be done in a manner so as to achieve good roads with minimum expenditure (Zore and Valunjkar, 2010). Economical flexible pavement structures require subgrade with good engineering properties to maximize the service life of roadway section and to minimize the thickness of flexible pavement structure considering strength, drainage, and ease of compaction (Sureka Naagesh and Sathyamurthy Sudhanva, 2013). An emerging trend of using waste material in soil stabilizing or soil strengthening is being operated all over the world in present times (Senadheera et al., 1998). The main reason behind this trend is the excessive production of waste, like fly ash, plastics, rice husk ash, which is not only hazardous but also creates deposition problems (Alemgena et al., 2012). There are two primary as well as empirical methodology approaches adopted for soil stabilization, namely mechanical and chemical stabilization (Ratna et al., 2013). These two methods of stabilization involve improvement of various engineering properties of soil through addition of chemicals or other materials to improve the existing soil (Jayakumar and Lau, 2012). This technique is generally cost effective. The road construction scenario has taken a big leap forward (Ziaie Moayed and Nazari, 2011). Still major hurdles, like constraint of funding, lack of good quality construction materials in the near vicinity, considerably challenge the road construction scenario which has taken a big leap forward (Praveen et al., 2006). So, the construction method should be such that, with minimum expenditure it is possible to have good roads. It is also a good idea to construct the roads in stages (Sinha et al., 2013). Fly ash roads can provide a far better surface than conventional roads due to the higher durability, which can be upgraded to a higher type of pavement at a later stage (Mallick and Mishra, 2011).

DOI: 10.1201/9781003217619-19

The present study tries to utilize and dispose these waste materials by reusing them in the construction of flexible pavements. Experiments are performed to determine the optimum combination of fly ash and lime for cost effective road construction, which is further used in preparing a generalized mathematical equation to find out the cost incurred. Reduction in the thickness layer is finally obtained making the construction of flexible pavement economical.

17.2 Methodology

Different laboratory tests in accordance with the project on stabilizing subgrade soil were conducted. Since subgrade refers to the in-situ soil where the stresses from the overlaying roadway will be distributed, so modification of its engineering properties for improved constructability, was precisely performed based on the following tests:

1. Wet sieve analysis
2. Determination of index properties of soil like liquid limit, plastic limit, and plasticity index
3. Modified proctor compaction test
4. CBR test.

Based on these tests our target is to evaluate the advantages in terms of Layer Thickness Reduction (LTR) after stabilizing the subgrade with different fly ash and lime content.

17.2.1 Fly Ash

When powdered coal is burnt in the furnace of thermal power plants, it produces a fine ash called fly ash, which is carried with the flue gases and is collected by using either cyclonic or electrostatic precipitators. About 80% of total ash produced is fly ash, the remaining ash sinters down to bottom of the furnace. This coarse fraction is referred to as bottom ash. In a wet disposal system, both fly ash and bottom ash are mixed and pumped to lagoons; this is called pump ash. Such huge quantity of waste is a problem that needs utmost attention. Utilizing this for road construction provides a valuable method for disposal. Also, as per the Govt. of India gazette notification dated 14.9.1999, it is mandatory that all the thermal power stations must make fly ash available for such purposes free of cost.

17.2.2 Use of Lime

Lime fly ash stabilized soil can be used for constructing subgrade of major roads. The use of stabilized fly ash subbase/base course is particularly attractive where fly ash is easily available and supplies of aggregates are unavailable or expensive. It is possible to construct a lime fly ash stabilized layer without admixing soil for a subbase layer. This would increase the utilization of fly ash and also prevent use of valuable topsoil.

17.2.3 SPSS and Cost Modeling

SPSS (Statistical Package for the Social Sciences) is a computer program used for survey authoring and deployment (IBM SPSS Data Collection), data mining (IBM SPSS Modeler), text analytics, statistical analysis, and collaboration and deployment (batch and automated scoring services). SPSS is among the most widely used programs for statistical analysis in social science. It is used by market researchers, health researchers, survey companies, government, education researchers, marketing organizations and others. In addition to statistical analysis, data management (case selection, file reshaping, creating derived data) and data documentation (a metadata dictionary is stored in the data file) are features of the base software. The generalized form of obtained equation is as follows: -

$$Y_p = a_0 + a_1 x_1 + a_2 x_2 + \ldots + a_n x_n$$

Development of Flexible Pavement Cost Models 141

where Y_p is the dependent variable, a_0 is the regression constant, $a_1, a_2 \ldots a_n$ are the regression coefficient and $x_1, x_2 \ldots x_n$ are the independent variables.

17.3 Specifications of IRC

17.3.1 Subgrade Soil

In roads, the top 500mm of the cutting or embankment at the formation level shall be considered as subgrade. The subgrade, whether it is in cut or fill, shall be compacted to utilize its inherent strength and prevent permanent deformation because of additional compaction by traffic. A minimum of 100% of modified proctor compaction should be attained in the top 500mm of the subgrade. For clay soil, the minimum compaction for subgrade should be 95%of modified proctor compaction and the compaction should be done at moisture content plus 2% OMC and dry density corresponding to modified proctor compaction and soaked in water for 4 days prior to testing. One or two CBR test should be done per km depending on the variation of the soil.

17.3.2 Liquid Limit

The permissible value of liquid limit is up to 70%.

17.3.3 Plasticity Index

The allowable value of plasticity index should be greater than 8% and is allowable up to 45% and the soil whose value of plasticity index is greater than 8% is suitable for stabilization.

17.3.4 Density Requirement

The maximum dry density should not be less than 1.75gm/cm^3.

17.3.5 CBR

The value of CBR should be equal to or greater than 2%. Table 17.1 shows the mechanical characteristics of the unstabilized soil that has been used for the experiment.

Lime and fly ash at different percentages were mixed with the soil and the variation of CBR values in first 4, 7, and 12 days has been depicted in Tables 17.2, 17.3, and 17.4 respectively. Figure 17.1 presents graphically the trend of increasing soil strength

TABLE 17.1

Properties of Unstabilized Soil

Property	Soil
Maximum dry density (kN/m³)	19.45
OMC (%)	13.30
CBR (%) (Heavy Compaction)	2.30
Liquid limit (%)	24.40
Plastic limit (%)	12.3
Plasticity index (%)	12
Unified soil classification	CL
Classified as per AASHTO	CL
Typical name	Inorganic clay of low plasticity

TABLE 17.2

Variation of 4-Day CBR Values with Both Lime and Fly Ash Content

Soil +lime +fly ashcontent	90% soil+ 10%fly ash +2% lime	90% soil+ 10%fly ash +3% lime	80% soil+ 20%fly ash +2% lime	80% soil+ 20%fly ash +3% lime	70% soil+ 30%fly ash +2% lime	70% soil+ 30%fly ash +3% lime	100%soil + 3%lime
4-day CBR (%)	17.54	18.6	21.26	31.66	30.53	36.53	22.3

TABLE 17.3

Variation of 7-Day CBR Values with both Lime and Fly Ash Content

Soil +lime +fly ashcontent	90% soil +10% fly ash + 2% lime	90% soil +10% fly ash + 3%lime	80% soil +20% fly ash + 2%lime	80% soil+ 20% fly ash + 3%lime	70% soil +30% fly ash + 2%lime	70% soil +30% fly ash + 3%lime	100%soil +3%lime
7-day CBR (%)	16.56	25.33	25.33	34.5	33.1	37.02	20.46

TABLE 17.4

Variation of 12-Day CBR Values with both Lime and Fly Ash Content

Soil +lime +fly ash content	90% soil +10% fly ash + 2%lime	90% soil +10% fly ash + 3%lime	80% soil +20% fly ash + 2%lime	80% soil+ 20% fly ash + 3%lime	70% soil +30% fly ash + 2%lime	70% soil +30% fly ash + 3%lime	100%soil +3%lime
12-day CBR (%)	30.0	35.2	35.8	49.7	48.0	55.8	60.8

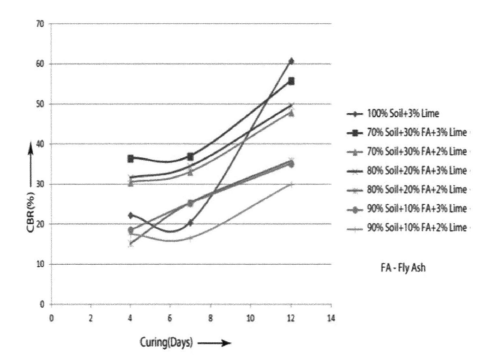

FIGURE 17.1 Graph showing CBR vs days of curing.

Development of Flexible Pavement Cost Models

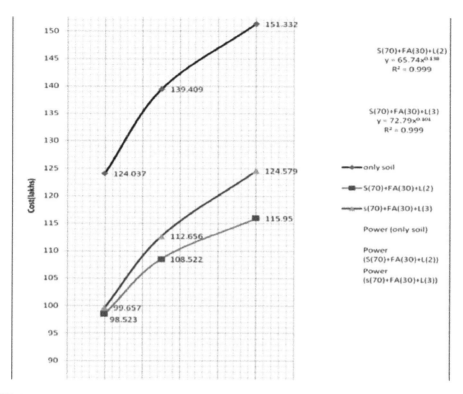

FIGURE 17.2 Variation of cost with respect to traffic volume in accordance with different combination of soil, fly ash & lime.

17.4 Economic Analysis

The main objective of this project is to evaluate the benefits in terms of Layer Reduction Thickness (LRT) on stabilizing the subgrade soil with lime and fly ash. The thickness of different layers of flexible pavement resting on stabilized and unstabilized subgrade for a traffic of 20 million standard axle (msa), 50 million standard axle (msa) and 100 million standard axle (msa) has been evaluated. The thickness of subgrade has been taken as 500mm. These model thicknesses are subsequently used for estimating the quantities and economics of a stabilized flexible pavement (Figure 17.2).

17.5 Discussion and Conclusion

- Durability of the subgrade is directly dependent on the CBR value. Hence, for a combination of 70:30:3, the maximum value of CBR is obtained that is 55.8%.
- It is recommended to use the above combination so as to maximize the use of fly ash thereby reducing the cost of construction of roads, saving valuable land use from filling of fly ash and less pollution hazard from storage of fly ash.
- A generalized equation using SPSS Software for cost model prediction was developed and the inter-relation between cost, CBR and traffic was known.

REFERENCES

Alemgena S., Marius, H., Molenaar, A.A., Houben, J.M. (2012), Investigation of the resilient behavior of granular base materials with simple test apparatus. *Materials and Structures,* 45(5), pp. 695–705.

IRC. (2012). *Guid*elines for the design of flexible pavements. Indian Code of Practice, IRC.

Jayakumar M., and Lau, C. (2012), Experimental studies on treated sub-base soil with fly-ash and cement for sustainable design recommendations world academy of science, *Engineering and Technology,* 68, pp. 611–614.

Mallick, S. R. and Mishra, M.K.(2011), *Replacement of Conventional Material with FCMs in Sub-Base of Opencast Mine Haul Road to Reduce Strain- An Investigation.* Department of Mining Engineering, NIT, Rourkela.

Praveen, Kumar, Chandra, S., and Vishal, R. (2006), Comparative study of different subbase materials. *Journal of material in Civil Engineering (ASCE)*, 18(4), pp. 576–580.

Ratna, Prasad R., Darga Kumar, N., and Janardhana, M. (2013), Effect of fly-ash on CBR and DCPT results of granular sub base subjected to heavy compaction. *International Journal of Scientific & Engineering Research*, 4(5), pp. 49–54.

Senadheera S., and Nash, P., and Rana, A. (1998), *Characterization of the Behavior of Granular Road Material Containing Glass Cullet.* Department of Civil Engineering, Texas Tech University, Lubbock, Texas, USA.

Sinha A.K., Havanagi, V.G., Ranjan, A., Mathur, S. (2013), Steel slag waste material for the construction of road. *Indian Highways,* 41 (10), pp. 1–12.

Sureka Naagesh, R., and Sathyamurthy Sudhanva, S. (2013), Laboratory studies on strength and bearing capacity of GSB-soil subgrade composites. *International Journals of Innovations in Engineering and Technology (IJIET)*, 2(2), pp. 245–254.

Ziaie Moayed, R., and Nazari, M. (2011), Effect of Utilization of Geo-synthetic on Reducing the Required Thickness of Sub-base Layer of a Two Layered Soil. *World Academy of Science, Engineering and Technology, Paris, France,* 49, (175), pp. 963–967.

Zore, T. D., and Valunjkar, S.S. (2010), Utilization of fly-ash and steel slag in road construction – A comparative study. *Electronic Journal of Geo-technical Engineering (EJGE)*,15.

18

Use of Fly Ash and Lime for Attainment of CN Properties in a Swelling Soil

K.V.N. Laxma Naik, A. Thanusree, and C. H. Sudha Rani
Sri Venkateswara University, Tirupati, India

CONTENTS

18.1 Introduction 145
18.2 Experimental Work 146
     18.2.1 Methods Adopted.. 147
18.3 Tests Results and Discussion 147
18.4 Liquid Limit 147
18.5 Compressive Strength... 148
18.6 Swelling Pressure @ OMC 149
18.7 Compression Index @ OMC... 150
18.8 Compression Index @ LL 150
18.9 Conclusions 151
References 151

18.1 Introduction

Often, project sites are located in areas with problematic soils. Depending on the nature of the project and nature of soil, the design solution may involve the expensive option of removal and replacement of the problematic soils. The replacement option typically entails use of crushed rock, gravel or lightweight aggregates, which implies higher cost and involves the use of limited natural resources. Alternately, ground improvement techniques such as stone columns, grouting, wick drains may be used to strengthen the soil to effect the desired changes in soil properties. Expansive soils are one of the problematic soils spread in arid and semi-arid regions of the globe. This paper focuses on studying the feasibility of using fly ash and/or lime to bring desirable changes in the swelling characteristics of mechanically stabilized expansive soil, specifically on index properties, compaction, shrinkage and swelling characteristics.

Mir (2015) reported results of tests conducted on expansive soil mixed with fly ash in different percentages by weight of dry soil varying from 10% to 80% and subjected to various tests. The important properties that are necessary for using fly ash in many geotechnical applications are index properties, compaction characteristics, compressibility characteristics, permeability and strength. Based on test results, it has been found that using fly ash for improvement of soils has a two-fold advantage. First, to avoid the tremendous environmental problems caused by large-scale dumping of fly ash and second, to reduce the cost of stabilization of problematic/marginal soils and improving their engineering properties for safe construction of engineering structures. Srinivas, Prasad & Rao (2016) reported a study on improvement of expansive soil by using CNS soil to improve existing expansive clay soil without using any admixtures, i.e., by using of CNS (Cohesive Non-Swelling) soils for improving a problematic soil so as to maintain environmental balance of existing soils. Studies on the plasticity and compaction characteristics of soil+CNS soil mixtures with additions of different percentages, cohesive non-swelling on selected expansive soil, proved to reduce the plasticity characteristics and improve the strength of

DOI: 10.1201/9781003217619-20

selected expansive soils (Sudha Rani, 2013; Rani & Suresh 2013). Reports on stabilizing expansive soils with admixtures like lime, cement, chemical, etc. has been found to be effective in improving their properties but, uniform blending of large quantities of soil with admixtures is difficult. Among the several methods adopted for improving the performance of expansive soils, provision of a stabilized cushion of fly ash was found to yield satisfactory results (Rao & Sridevi 2011). Cohesive CNS is a phenomenon arising out of combination of Coulombian and Newtonian forces. Any system of material capable of developing this phenomenon may be designated as CNS (Radhakrishnan et al., 2016). Both Newtonian and Coulombian forces alter H-O-H in the Montmorillonite structure and possess high shear strength. The CNS technology is for construction on expansive soils. Results from small test predictions are in considerable variance with field reality: as such tests are not properly simulated to account for exponential variations. The objective of the present study is to find the optimum percentage or dosage of admixtures to the selected soil to represent CNS material (Masoud, 2015; Katti et al., 2010).

18.2 Experimental Work

Materials Used: The details pertaining to soil, fly ash used in this investigation are given in the Tables 18.1 and 18.2 respectively. The soils used in the present investigation were obtained from the Tirupati Airport region. The required amount of soil was collected from the trail pit at a different depth of 2.5m below the ground level, since the top soil is likely to contain organic matter and other foreign materials.

TABLE 18.1

Physical Properties of Fly Ash

Characteristics	Value
Specific Gravity	2.023
Bulk Density (kN/m^3)	14.50
Color	Gray
OMC (%)	21.0
MDD (kN/m^3)	11.98

TABLE 18.2

Properties of the Soil

S. No	Experiment	Value
1	Specific Gravity	2.71
2	Liquid Limit (%)	79.6
3	Plastic Limit (%)	NP
4	Free Swell Index	220.00
5	Coarse Fraction (%)	77.4
6	Fine Fraction (%)	22.6
7	IS Soil classification	SC (Clay Sand)
8	Maximum Dry Density (kN/m^3)	18.50
9	Optimum Moisture Content (%)	12.0
10	Unconfined compressive Strength (kPa)	124.00
11	Cohesion (kPa)	62.00
12	Swelling Pressure @ OMC (kPa)	180.00
13	Compression Index @ OMC	0.311
14	Compression Index @ L	0.415

Use of Fly Ash and Lime for Attainment

The soil is classified as 'SC' as per IS classification (IS 1498:1978).

18.2.1 Methods Adopted

The following tests were conducted on soil samples in the presented investigation:

1. Liquid limit test
2. Plastic limit test
3. Differential free swell index test
4. Compaction test
5. Unconfined Compressive Strength test
6. Consolidation test

18.3 Tests Results and Discussion

As soil samples possess liquid limit (LL) more than 50% and have a very high degree of swelling, suitable proportions of fly ash are added for reducing the LL to the desired value. Since the LL range specified by Prof. R.K. Katti was not found even with addition of 60% of fly ash by weight of soil, so, (fly ash + lime) were added in combination. With the increase in percentage of (fly ash + lime) to the soil the LL was found to decrease. Specified range for suitability of soil as CNS material proposed by Prof. R.K. Katti was found with (20% fly ash + 5% lime) added to the soil. Further tests (Unconfined Compression tests, swelling tests and consolidation tests) are conducted by addition of (20% fly ash + 5% lime) to the soil. Compression index is determined at two water contents, i.e., at Optimum Moisture Content, and at liquid limit, in order to assess maximum compressibility case. Table 18.3 shows the experimental results of soil alone and soil admixed with (20% fly ash + 5% lime).

18.4 Liquid Limit

The variation of LL of soil with addition of fly ash alone and, (fly ash + lime) in combination are shown in Figures 18.1 and 18.2 respectively. From Figure 18.1 it can be observed that the LL of soil is decreasing with increase in percentage of fly ash but, as per Prof. R.K. Katti, the LL for CNS soil should be between 30% and 50%, as it is not achieved even up to 60% of soil, so 5% lime is added along with fly ash.

TABLE 18.3

Test Results on Soil and (Soil +20% Fly Ash +5% Lime) Mixtures

		Value		%Increase/ %Decrease with reference to soil
S.No	Property	Soil	Soil+20% fly-ash+5% lime	
1	Specific Gravity	2.71	2.65	2.26↓
2	Liquid Limit (%)	79.60	46.00	73.04↓
3	Free Swell Index (%)	220.00	100.00	1.20↓
4	Maximum Dry Density (kN/m³)	18.50	18.30	1.09↓
5	Optimum Moisture Content (%)	12.00	14.00	14.28↑
6	Unconfined Strength (kPa)	129.00	179.00	27.93↑
7	Cohesion (kPa)	73.00	89.50	18.43↑
8	Swelling pressure @ OMC (kPa)	180.00	10.00	1700↓
9	Compression Index @ OMC	0.311	0.194	60.30↓
10	Compression Index @ LL	0.415	0.222	86.93↓

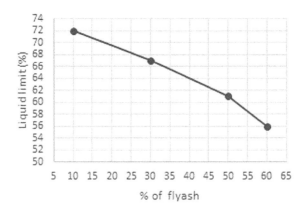

FIGURE 18.1 Variation of LL with addition of different percentages of fly ash to soil.

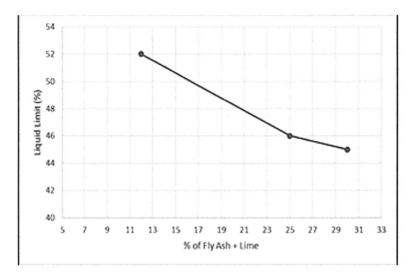

FIGURE 18.2 Variation of LL with varying proportions of (fly ash + lime) to soil.

From Figure 18.2 it can be observed that the LL of soil is decreasing with the addition of fly ash and lime in combination. Maximum reduction of LL can be observed for (25% fly ash + 5% lime) but, further, the addition is found to reduce below 30% value and does not represent required CNS material range; further incremental percentage addition of (fly ash + lime) was terminated. The decrement in LL of soil (fly ash + lime) mixtures with (25% flyash+ 5% lime) is nominal when compared to (20% fly ash + 5% lime) soil mixture so, further tests are conducted on soil-(20% fly ash + 5% lime) mixtures.

18.5 Compressive Strength

The Unconfined Compression test results for soil alone and for soil with addition of (fly ash + lime) in combinations compacted at OMC and MDD are presented in Figure 18.3. From Figure 18.3 it can be observed that the compressive strength increases with the addition of 20% fly ash + 5% lime to the soil by 30%.

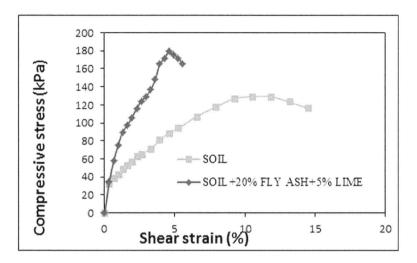

FIGURE 18.3 Compressive stress vs strain curves for soil alone and (soil + 20% fly ash + 5% lime) mixture.

18.6 Swelling Pressure @ OMC

The results of swelling pressure tests, where load versus change in thickness of soil samples for soil alone, and soil with addition of (fly ash + lime) in combinations are represented in Figure 18.4. The swelling pressure can be observed to decrease drastically from 180kPa to 10kPa with addition of (20% fly ash +5% lime) when compared to soil alone. This property can be given highest preference as a representative of CNS material.

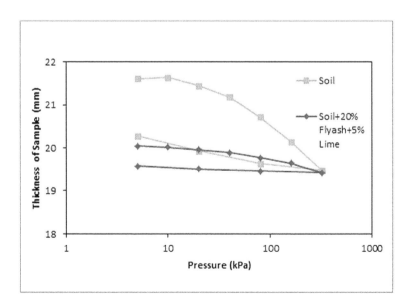

FIGURE 18.4 Thickness of sample vs Log σ curves for soil alone and (soil + 20% fly ash + 5% lime) mixture.

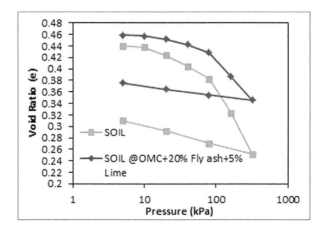

FIGURE 18.5 Void ratio (e) vs log p curves for soil alone and (soil + 20% fly ash + 5% lime) mixture compacted @ OMC.

18.7 Compression Index @ OMC

Void ratio (e) vs log P curves are presented in Figure 18.5, based on consolidation test results of soil alone and (soil+ 20% fly ash +5% lime) mixtures compacted at OMC and MDD. The slope of straight-line portions represents the respective compression index. The compression index can be observed to decrease by 60% with the addition of 20% of fly ash + 5% lime to soil.

18.8 Compression Index @ LL

Consolidation test results of soil alone and (soil+ 20% fly ash +5% lime) mixtures compacted at LL water content are presented in Figure 18.6. It can be observed that the compression index decreased by 90% with the addition of (20% of fly ash + 5% lime) to soil even at LL water content.

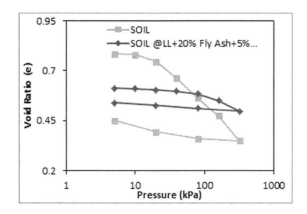

FIGURE 18.6 Void ratio (e) vs log P curves for soil alone and (soil + 20% fly ash + 5% lime) mixture compacted @ LL.

18.9 Conclusions

The following concluding remarks are made from the present work:

1. The liquid limit of soil decreased with addition of 20% fly ash and 5% lime from 79.60% (original soil) to 46% (the decrement is about 73.04% compared to the original soil), satisfying the LL range values recommended by Prof. R.K. Katti, as a CNS material.
2. The Unconfined Compressive Strength increased from 129kPa (original soil) to 179kPa of soil with addition of (20% fly ash +5% lime).
3. The swelling pressure of soil mixed with (20% fly ash +5% lime) decreased from 180kPa (original soil) to 10kPa (satisfying R.K Katti specifications.
4. The compression index of soil compacted with LL water content decreased from 0.415 (original soil) to 0.222 with addition of (20% fly ash + 5% lime). (the decrement is about 60.30% compared to the original soil). The compression index of soil @ OMC decreased with addition of (20% fly ash + 5% lime) by 60.30% when compared to the original soil.
5. The (20% fly ash + 5% lime) combination can be used for selected expansive soil for representing CNS material properties reducing environmental pollution and exploiting natural resources.

REFERENCES

Kola Srinivas, Prasad, D.S.V., and Rao, V.K.L. (2016) "A study on improvement of expansive soil by using CNS". *International Journal of Innovative Research in Technology*, Volume 3, Issue 3, pp. 37–56.

Talal Y. Masoud. (2015) "Non-dimension chart to determine the thickness of CNS soil to minimize the effect of expansive Soil exerted on Circular Footing". *4th International Conference on Chemical, Ecology and Environmental Sciences (ICEES'2015 Pattaya)* (Thailand).

Sudha Rani, C.H. (2013) "Investigation on engineering properties of soil-mixtures comprising of expansive soils and a cohesive non-swelling soil". *International Journal of Engineering Research and Applications (IJERA)*, Volume 3, Issue 3, pp. 155–160.

Katti, R., Kulkarni, U., Katti, A., and Kulkarni, R. (2010) "Stabilization of embankment on expansive soil" *International Conference Experimental and Applied Modeling of Unsaturated Soils*, Shanghai, China, June 3–5, 2010.

Radhakrishnan, G., Anjan Kumar, M., and Prasada Raj, G.V.R. (2016) "Swelling properties of expansive soils treated with chemicals and fly ash". *American Journal of Engineering Research*, Volume 3, issue 4, pp. 312–331.

19

Interface Shear Strengths between Bagasse Ash and Geogrid

Aditya Kumar Bhoi, Jnanendra Nath Mandal, and Ashish Juneja
IIT Bombay, Mumbai, India

CONTENTS

19.1 Introduction 153
19.2 Testing Materials 154
      19.2.1 Bagasse Ash 154
19.3 Geogrid 155
19.4 Testing Methods 155
19.5 Results and Discussion... 156
      19.5.1 Direct Shear Test Results 156
19.6 Interface Shear Test Results 157
19.7 Friction Efficiency Factors (E_Φ) 158
19.8 Conclusion... 158
Acknowledgment 159
References.. 159

19.1 Introduction

India is the second largest sugar cane producer in world with an average of 335 million tonne of sugar cane per year from the last ten year (FAO, 2019). Sugar industries produces 32–35% of bagasse or about 107–117 million tonnes of bagasse per year, which is generally used as fuel in some industries to produce steam and electricity to meet their own demand (The Sugar Technologists Association of India., 1959). Burning of bagasse produces bagasse ash. The amount of bagasse ash is about 0.3% of the weight of cane (Lee et al., 1965). This leads to a generation of about 1 million tonnes of bagasse ash per year. This bagasse ash can be classified into two subgroups, i.e., the first one is bottom ash, which used to accumulate in the bottom of boiler's combustion chamber, and the second one is fly ash, which is collected from the particulate emission control unit. Bagasse ash is generally dumped in the sugar cane field (Deepika et al., 2017). The open-dumping causes environmental pollution; the waste needs to be managed in a sustainable manner for civil engineering construction. Earlier studies reported the incorporation of bagasse ash in cement concrete as a pozzolanic material (Bahurudeen et al., 2014; Bahurudeen et al., 2015; Chusilp et al., 2009; Cordeiro et al., 2009; Fairbairn et al., 2010; Frías et al., 2011; Ganesan et al., 2007; Rajasekar, 2018; Rukzon & Chindaprasirt, 2012; Shafiq et al., 2016; Singh et al., 2000; Sua-Iam & Makul, 2013; Villar-Cocina et al., 2013). Some scholars have used the bagasse ash along with cement to stabilize expansive soil (Kumar Yadav et al., 2017; Sabat, 2012; Osinubi et al., 2009; Jamsawang et al., 2017). Bagasse ash has physical properties similar to sand, with high silicon content and crystalline structure. As a result of which many scholars are exploring the feasibility of bagasse ash as a fine aggregate in cement concrete (Aigbodion et al., 2010; Almeida et al., 2015; Arif et al., 2016; Bilir et al., 2015; Moretti et al., 2016; Prusty et al., 2016; Purohit & Michaelowa, 2007; Rashad, 2016; Sales & Lima, 2010; Shafigh et al., 2014).

DOI: 10.1201/9781003217619-21

Most of the literature shows that many scholars are working with bagasse ash to assess its usability as a pozzolanic material in concrete, as a stabilizer for expansive soil, and as a fine aggregate for cement concrete. However, there are no research works related to bagasse ash as a backfill material and its interaction with geogrid. The current research focuses on the potential use of bagasse ash as a backfill material and its interaction with geogrid.

19.2 Testing Materials

19.2.1 Bagasse Ash

The bagasse ash used in this study was collected from Sahyadri Sahakari Sakhar Karkhana Ltd. situated in Yashvantnagar, District. Satara, Maharashtra, India. The bottom ash was collected from the bagasse ash dump yard, whereas the fly ash was collected from the fly ash discharging chute. Both samples were oven-dried in the laboratory before any testing. The bottom ash and fly ash from now will be referred to as BA and FA, respectively. The specific gravity of bottom ash and fly ash were found out to be 2.37 and 2.13 as per IS:2720 (IS:2720 (Part III/Sec 1) 1980, 2002). The grain size distribution of the bottom ash and fly ash were carried out by sieving and hydrometer analysis according to IS:2720 (IS:2720 (Part IV) 1985, 2006) and are depicted in Figure 19.1. Atterberg's limits were carried out to determine the plastic limit according to IS: 2720 (IS:2720 (Part V) 1985, 2006). But the authors were unable to measure the liquid limit of bottom ash and fly ash, as they started bleeding during the test. Similarly, authors were unable to make 3.2mm thread by hand-rolling for the plastic limit. Hence, the bottom ash and fly ash were identified as non-plastic soil (IS:2720 (Part V) 1985, 2006). Bottom ash and fly ash were classified as inorganic silt (ML), according to Indian standard IS:1498-1970 (IS:1498-1970, 2007). The standard Proctor test was conducted as per IS:2720 (IS:2720 (Part VII) 1980, 2011) and are depicted in Figure 19.2. The maximum dry density and optimum moisture content (OMC) of bottom ash were found

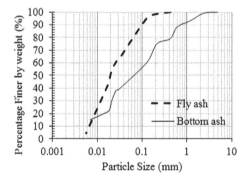

FIGURE 19.1 Grain size distribution curve of bagasse ash.

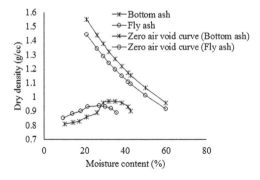

FIGURE 19.2 Compaction curve of bagasse ash.

Interface Shear Strengths

out to be 0.97 g/cm³ and 32%. Similarly, the maximum dry density and OMC of fly ash were found out to be 0.94 g/cm³ and 27%.

19.3 Geogrid

The opening size of the geogrid should be 10 times more than the average particle size of backfill material to achieve good bondage between soil and geogrid (Izawa & Kuwano, 2010). Hence a model geogrid with opening size 1.37mm is used in this study. The mass per unit area per ASTM D D5261–10 (ASTM D5261–10, 2018) of geogrid as was found out to be 115.7g/m². The thickness of geogrid was found out to be 0.3mm as per ASTM D5199–12 (ASTM D5199–12, 2019). The tensile strength was found out to be 1.57kN/m in machine direction (ASTM D4595–17, 2017).

19.4 Testing Methods

Laboratory investigation was conducted using a direct shear device of size 60mm × 60mm, according to IS:2720 (IS:2720 (Part 13) 1986, 2002). The entire laboratory programme was divided into two groups. The first group was designed to determine a bagasse ash friction angle. In this group, friction angle of bottom ash and fly ash were determined separately. The second group was designed to determine the friction angle of the bagasse ash–geogrid interface. Hence, bottom ash-geogrid interface (BA/G) and fly ash-geogrid interface (FA/G) were tested separately. The test samples were prepared at 90% of the maximum dry density to maintain uniformity among the sample. The cross-sectional schematic view of bottom ash to bottom ash (or fly ash to fly ash) and bottom ash to geogrid (or fly ash to geogrid) samples in direct shear box are shown in Figure 19.3.

The first group was designed to determine the bagasse ash (i.e., bottom ash and fly ash) friction angle. The shear box was filled with a predetermined amount of bottom ash (or fly ash. Then the sample was compacted by using a hammer in several layers to get a density corresponding to 90% of the maximum dry density. The samples were sheared under three different normal stresses (i.e., 50kPa, 100kPa, and 150kPa). The bottom ash samples sheared under 50kPa, 100kPa, and 150kPa normal stresses were identified as 50(BA), 100(BA) and 150(BA), respectively. Similarly, the fly ash samples sheared under 50kPa, 100kPa, and 150kPa normal stresses were identified as 50(FA), 100(FA) and 150(FA), respectively.

The second group of laboratory test was designed to determine the interface friction angle of bagasse ash–geogrid. Hence, bottom ash-geogrid interface (BA/G) and fly ash-geogrid interface (FA/G) were tested separately. A spacer block, warped with geogrid on its top was placed in the lower half of direct shear box (see Figure 19.4(a)). Then, the upper half of the shear box was placed and fixed over the lower half of shear box, followed by placement of predetermined amount of bottom ash (or fly ash) in several layers and compacted by using a hammer (see Figure 19.4(b)). Then the samples were sheared under 50kPa, 100kPa, and 150kPa. The bottom ash-geogrid samples sheared under 50kPa, 100kPa, and 150kPa normal stresses were identified as 50(BA/G), 100(BA/G) and 150(BA/G), respectively. Similarly, the fly ash–geogrid samples sheared under 50kPa, 100kPa, and 150kPa normal stresses were identified as 50(FA/G), 100(FA/G) and 150(FA/G) respectively.

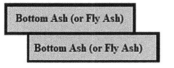
(a) Bottom ash to Bottom ash [or Fly ash to Fly ash]

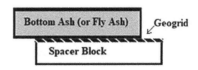
(b) Bottom ash to Geogrid [or Fly ash to Geogrid]

FIGURE 19.3 (a, b) Schematic diagrams of direct shear tests.

FIGURE 19.4 Interface shear test specimen arrangement in the shear box.

All the sample were prepared in such a manner that there was no friction between the lower half and upper half of direct shear test device. To achieve this goal, a gap of 1mm between lower half and upper half of shear box was maintained for all the samples. The upper half of direct shear device remained constant, whereas the lower half of direct shear device was pushed at a speed of 1mm/min. The horizontal displacement and shear force were measured by using dial gauges and a proving ring, respectively, during shearing.

19.5 Results and Discussion

19.5.1 Direct Shear Test Results

Figure 19.5a, represent the variation of shear stress of bottom ash and fly ash with different horizontal displacement, obtained from direct shear testing. From this figure it is observed that both bottom ash and fly ash exhibit distinct peak shear stress when sheared under higher normal stress, and an ultimate shear stress when sheared under lower normal stress. Peak shear stress was considered as a failure condition

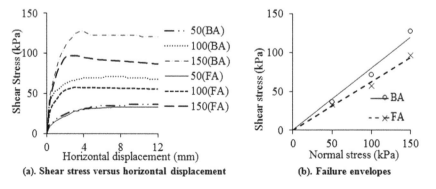

FIGURE 19.5 Direct shear test of bottom ash and fly ash.

Interface Shear Strengths 157

for the sample with distinct peak shear stress. Whereas an initial horizontal tangent to the shear stress-horizontal displacement curve was considered as failure condition for sample with an ultimate stress (Bareither et al., 2008a; Bareither et al., 2008b). The bottom ash fails at 6mm horizontal displacement, with a shear stress of 36kPa, when sheared under a normal stress of 50kPa. Similarly, bottom ash fails at 6.3mm horizontal displacement, with a shear stress of 71kPa, when sheared under 100kPa normal stress. And the peak shear stress and horizontal displacement are 128kPa and 3.9mm, respectively, when sheared under 150kPa normal stress. The shear stress to horizontal displacement curve of fly ash under 50kPa normal stress is similar to bottom ash, but the ploughing effect is not visible in this case. The fly ash fails at 3mm horizontal displacement and 57kPa, and 2.7mm horizontal displacement and 97kPa, when sheared under 100kPa and 15 kPa normal stress.

Figure 19.5b, represent Mohr-Coulomb failure envelopes of bottom ash and fly ash, obtained from direct shear testing. Linear least squares regression with a non-negative intercept was used here to obtain the failure envelope (Bareither et al., 2008a; Bareither et al., 2008b). The friction angle of bottom ash is found out to be 41° with a coefficient of determination (R^2) value of 0.965. Similarly, the friction angle of fly ash is found out to be 32° with a coefficient of determination (R^2) value of 0.979. The main shear strength parameter of bagasse ash is friction angle. The bottom ash friction angle is more than the fly ash friction angle, because bottom ash has less finer compared to fly ash.

19.6 Interface Shear Test Results

The shear stress-horizontal displacement curve of bottom ash-geogrid interface and pure bottom ash are shown in Figure 19.6a. From this figure, it is clear that shear stress–horizontal displacement curve of the BA/G interface is similar to the shear stress–horizontal displacement curve of BA, when sheared under a normal stress of 50kPa and 100kPa. But the BA/G interface is only able to reach 96kPa shear stress at a horizontal displacement of 5.1mm. This is quite low compared to BA. Figure 19.7a, shows the shear stress–horizontal displacement curve of FA/G interface along with BA. From the Figure 19.7a, it is clear that shear stress of FA/G interface is low compared to shear stress of BA, when sheared under similar normal stress. The shear stresses are 22kPa, 48kPa and 66kPa at failure, when sheared under 50kPa, 100kPa and 150kPa normal stress. But the FA/G and BA/G interfaces fail at higher magnitude of horizontal displacement compared to only FA and BA sample.

Figure 19.6b, shows the Mohr-Coulomb failure envelopes of the BA/G interface along with BA. The friction angle of the BA/G interface is found out to be 33° with a coefficient of determination (R^2) value of 0.994. Similarly, Figure 19.7b, shows the Mohr-Coulomb failure envelopes of the FA/G interface along with FA. The friction angle of FA/G interface is found out to be 25° with a coefficient of determination (R^2) value of 0.988.

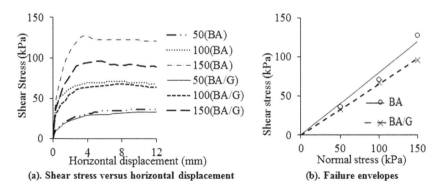

FIGURE 19.6 Interface shear test of bottom ash and geogrid.

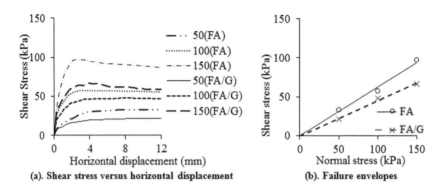

FIGURE 19.7 Interface shear test of fly ash and geogrid.

TABLE 19.1

Internal Friction Angle (Φ), Interfacial Friction Angle (δ_G), and Friction Efficiency Factors (EΦ) Values for Bottom Ash/Geogrid Interface and Fly Ash/Geogrid Interface

Bagasse ash type	Φ(°)	δ_G (°)	$(E_\Phi) = (\tan(\delta_G))/(\tan(\Phi))$
Bottom ash	41	33	0.75
Fly ash	32	25	0.74

19.7 Friction Efficiency Factors (E_Φ)

The friction efficiency factors (E_Φ) between bagasse ash and geogrid using direct shear test were evaluated as follows:

$$\text{Friction efficiency factors}(E_\Phi) = (\tan(\delta_G))/(\tan(\Phi)) \tag{19.1}$$

Where, δ_G is interfacial friction angle between bagasse ash and geogrid and Φ is friction angle of bagasse ash at failure.

The friction angle (Φ) of bagasse ash at failure and interfacial friction angle (δ_G) at failure were evaluated for bottom ash and fly ash with geogrid are presented in Table 19.1. The friction efficiency factors (E_Φ) for both BA/G and FA/G interface are below 1. The amount of shear resistance developed between bagasse ash particles during direct shear test is a function of interlocking between particles, whereas the magnitude of shear resistance developed between bagasse ash particles and the surface of geogrid during direct shear test is mainly a function of contact area between bagasse ash particles and geogrid. In general, the friction efficiency factors (E_Φ) are lower than 1 (Cazzuffi et al., 1993; Han et al., 2018; Liu et al., 2009; Mosallanezhad et al., 2016).

19.8 Conclusion

The present study deals with the experimental analysis of bagasse ash–geogrid interface. Considering the direct shear tests on bottom ash and fly ash, and interface shear testing between bagasse ash and geogrid, the following conclusions are drawn from this present study.

1. The bottom ash friction angle is more than the fly ash friction angle.
2. Bottom ash exhibit ploughing behaviour when sheared under lower normal stress.

Interface Shear Strengths

3. Interface shear resistance mainly depend on the contact area between bagasse ash particles and geogrid.
4. The friction efficiency factors (E_Φ) of BA/G interface and FA/G interface are nearly the same. This indicates particle size does not affect the interface shear strength.

ACKNOWLEDGMENT

Aditya Kumar Bhoi is thankful to the facilities received from AICTE under QIP scheme for carrying out this project.

REFERENCES

Aigbodion, V.S., Hassan, S.B., Ause, T., and Nyior, G.B., 2010. Potential utilization of solid waste (bagasse ash). *Journal of Minerals & Materials Characterization & Engineering*, 9(1): 67–77.

Almeida, F.C.R., Sales, A., Moretti, J.P., and Mendes, P.C.D., 2015. Sugarcane bagasse ash sand (SBAS): Brazilian agroindustrial by-product for use in mortar. *Construction and Building Materials*, 82: 31–38.

Arif, E., Clark, M.W., and Lake, N., 2016. Sugar cane bagasse ash from a high efficiency co-generation boiler: Applications in cement and mortar production. *Construction and Building Materials*, Elsevier Ltd, 128: 287–297.

ASTM D4595–17, 2017. *Standard Test Method for Tensile Properties of Geotextiles by the Wide-Width Strip Method*. ASTM International, West Conshohocken, United States.

ASTM D5199–12, 2019. *Standard Test Method for Measuring the Nominal Thickness of Geosynthetics*. ASTM International, West Conshohocken, United States.

ASTM D5261–10, 2018. *Standard Test Method for Measuring Mass per Unit Area of Geotextiles*. ASTM International, West Conshohocken, United States.

Bahurudeen, A., Marckson, A.V., Kishore, A., and Santhanam, M., 2014. Development of sugarcane bagasse ash based Portland pozzolana cement and evaluation of compatibility with superplasticizers. *Construction and Building Materials*, Elsevier Ltd, 68: 465–475.

Bahurudeen, A., Kanraj, D., Gokul Dev, V., and Santhanam, M., 2015. Performance evaluation of sugarcane bagasse ash blended cement in concrete. *Cement and Concrete Composites*, Elsevier Ltd, 59: 77–88.

Bareither, C.A., Edil, T.B., Benson, C.H., and Mickelson, D.M. 2008a. Geological and Physical Factors Affecting the Friction Angle of Compacted Sands. *Journal of Geotechnical and Geoenvironmental Engineering*, 134(10): 1476–1489.

Bareither, C.A., Benson, C.H., and Edil, T.B., 2008b. Comparison of shear strength of sand backfills measured in small – scale and large – scale direct shear tests. *Canadian Geotechnical Journal*, 45: 1224–1236.

Bilir, T., Gencel, O., and Topcu, I.B., 2015. Properties of mortars with fly ash as fine aggregate. *Construction and Building Materials*, Elsevier Ltd, 93: 782–789.

Cazzuffi, D., Picarelli, L., Ricciuti, A., and Rimoldi, P., 1993. Laboratory investigations on the shear strength of geogrid reinforced soils, *ASTM Special Technical Publication* 1190: 119–137.

Chusilp, N., Jaturapitakkul, C., and Kiattikomol, K., 2009. Utilization of bagasse ash as a pozzolanic material in concrete. *Construction and Building Materials*, Elsevier Ltd, 23(11): 3352–3358.

Cordeiro, G.C., Toledo Filho, R.D., Tavares, L.M., and Fairbairn, E. de M.R., 2009. Ultrafine grinding of sugar cane bagasse ash for application as pozzolanic admixture in concrete. *Cement and Concrete Research*, Elsevier Ltd, 39(2): 110–115.

Deepika, S., Anand, G., Bahurudeen, A., and Santhanam, M., 2017. Construction products with sugarcane bagasse ash binder. *Journal of Materials in Civil Engineering*, 29(10): 1–10.

Fairbairn, E.M.R., Americano, B.B., Cordeiro, G.C., Paula, T.P., Toledo Filho, R.D., and Silvoso, M.M., 2010. Cement replacement by sugar cane bagasse ash: CO2 emissions reduction and potential for carbon credits. *Journal of Environmental Management*, Elsevier Ltd, 91(9): 1864–1871.

FAO, 2019. Crops. http://www.fao.org/faostat/en/#data/QC (Accessed July 10, 2019).

Frías, M., Villar, E., and Savastano, H., 2011. Brazilian sugar cane bagasse ashes from the cogeneration industry as active pozzolans for cement manufacture. *Cement and Concrete Composites*, 33(4): 490–496.

Ganesan, K., Rajagopal, K., and Thangavel, K., 2007. Evaluation of bagasse ash as supplementary cementitious material. *Cement and Concrete Composites*, 29(6): 515–524.

Han, B., Ling, J., Shu, X., Gong, H., and Huang, B., 2018. Durability of innovative construction materials and structures laboratory investigation of particle size effects on the shear behavior of aggregate-geogrid interface. *Construction and Building Materials*. 158: 1015–1025.

IS:1498-1970, 2007. *Classification and Identification of Soils for General Engineering Purposes*. Bureau of Indian Standards, New Delhi.

IS:2720 (Part III/Sec 1) 1980, 2002. *Determination of Specific Gravity: Fine Grained Soils*. Bureau of Indian Standards, New Delhi.

IS:2720 (Part IV) 1985, 2006. *Grain Size Analysis*. Bureau of Indian Standards, New Delhi.

IS:2720 (Part V) 1985, 2006. *Determination of Liquid and Plastic Limit*. Bureau of Indian Standards, New Delhi.

IS:2720 (Part VII) 1980, 2011. *Determination of Water Content-Dry Density Relation Using Light Compaction*. Bureau of Indian Standards, New Delhi.

IS:2720 (Part 13) 1986, 2002. *Direct Shear Test*. Bureau of Indian Standards, New Delhi.

Izawa, J., and Kuwano, J., 2010. Centrifuge modelling of geogrid reinforced soil walls subjected to pseudo-static loading. *International Journal of Physical Modelling in Geotechnics*, 10(1): 1–18.

Jamsawang, P., Poorahong, H., Yoobanpot, N., Songpiriyakij, S., and Jongpradist, P., 2017. Improvement of soft clay with cement and bagasse ash waste. *Construction and Building Materials*, 154: 61–71.

Kumar Yadav, A., Gaurav, K., Kishor, R., and Suman, S. K., 2017. Stabilization of alluvial soil for subgrade using rice husk ash, sugarcane bagasse ash and cow dung ash for rural roads. *International Journal of Pavement Research and Technology, Chinese Society of Pavement Engineering*, 10(3): 254–261.

Lee, L.H., Chan, Y.S., and Lio, S.T., 1965. The application of bagasse furnace ash to sugarcane fields. *Annual Report of Taiwan Sugar Expert Station*, 38: 53–79.

Liu, C.N., Ho, Y.H., and Huang, J.W., 2009. Large scale direct shear tests of soil/PET-yarn geogrid interfaces, *Geotextiles and Geomembranes*, 27: 19–30.

Moretti, J.P., Sales, A., Almeida, F.C.R., Rezende, M.A.M., and Gromboni, P.P., 2016. Joint use of construction waste (CW) and sugarcane bagasse ash sand (SBAS) in concrete. *Construction and Building Materials*, Elsevier Ltd, 113: 317–323.

Mosallanezhad, M., Alfaro, M.C., Hataf, N., and Taghavi, S.H.S., 2016. Performance of the new reinforcement system in the increase of shear strength of typical geogrid interface with soil. *Geotextiles and Geomembranes*, 44: 457–462.

Osinubi K.J., Bafyau V., and Eberemu A.O., 2009. Bagasse ash stabilization of lateritic soil. In: Yanful E.K. (ed.) *Appropriate Technologies for Environmental Protection in the Developing World*. Springer, Dordrecht: 271–280.

Prusty, J.K., Patro, S.K., and Basarkar, S.S., 2016. Concrete using agro-waste as fine aggregate for sustainable built environment – A review. *International Journal of Sustainable Built Environment, The Gulf Organisation for Research and Development*, Elsevier Ltd, 5: 312–333.

Purohit, P., and Michaelowa, A., 2007. CDM potential of bagasse cogeneration in India. *Energy Policy*, 35(10): 4779–4798.

Rajasekar, A., Arunachalam, K., Kottaisamy, M., and Saraswathy, V., 2018. Durability characteristics of ultra high strength concrete with treated sugarcane bagasse ash. *Construction and Building Materials*, Elsevier Ltd, 171: 350–356.

Rashad, A., 2016. Cementitious materials and agricultural wastes as natural fine aggregate replacement in conventional mortar and concrete. *Journal of Building Engineering*, Elsevier Ltd, 5: 119–141.

Rukzon, S., and Chindaprasirt, P., 2012. Utilization of bagasse ash in high strength concrete. *Materials and Design*, Elsevier Ltd, 34: 45–50.

Sabat, A.K., 2012. Utilization of bagasse ash and lime sludge for construction of flexible pavements in expansive soil areas. *Electronic Journal of Geotechnical Engineering*, 17 H: 1037–1046.

Sales, A., and Lima, S.A., 2010. Use of Brazilian sugarcane bagasse ash in concrete as sand replacement. *Waste Management*, Elsevier Ltd, 30(6): 1114–1122.

Shafigh, P., Mahmud, H. Bin, Jumaat, M. Z., and Zargar, M., 2014. Agricultural wastes as aggregate in concrete mixtures - A review." *Construction and Building Materials*, Elsevier Ltd, 53: 110–117.

Shafiq, N., Hussein, A.A.E., Nuruddin, M.F., and Mattarneh, H.Al., 2016. Effects of sugarcane bagasse ash on the properties of concrete. *Proceedings of the Institution of Civil Engineers – Engineering Sustainability*: 1–10.

Singh, N.B., Singh, V.D., and Rai, S., 2000. Hydration of bagasse ash-blended portland cement. *Cement and Concrete Research*, 30(9): 1485–1488.

Sua-Iam, G., and Makul, N., 2013. Use of increasing amounts of bagasse ash waste to produce self-compacting concrete by adding limestone powder waste. *Journal of Cleaner Production*, Elsevier Ltd, 57: 308–319.

The Sugar Technologists Association of India, 1959. *Indian Sugar Manual*. The Sugar Technologists Association of India, Kanpur, India.

Villar-Cocina, E., Frias, M., Hernandez-Ruiz, J., and Savastano, Jr. H., 2013. Pozzolanic behaviour of a bagasse ash from the boiler of a cuban sugar factory. *Advances in Cement Research*, 25(3): 136–142.

Part III

Sustainable Green Concrete

20 Experimental Investigation on Geopolymer Concrete with Low-Density Aggregate

Sanghamitra Jena, Ramakanta Panigrahi, and Subrat Kumar Padhy
Veer Surendra Sai University of Technology, India

CONTENTS

20.1 Introduction 165
20.2 Materials and Method 166
    20.2.1 Fly Ash 166
    20.2.2 Fine Aggregates 167
    20.2.3 Coarse Aggregate 167
    20.2.4 Low-Density Aggregate 167
20.3 Alkaline Solutions 169
20.4 Mixing, Casting and Curing 169
20.5 Result and Discussion 170
    20.5.1 Workability 170
    20.5.2 Density (ρ) 170
    20.5.3 Compressive Strength (CS) 170
    20.5.4 Split Tensile Strength 171
    20.5.5 Flexural Strength 171
20.6 Conclusion 172
References 172

20.1 Introduction

Concrete is the most widely used material for construction. In this, cement is the main ingredient. Nowadays the demand for cement is increasing rapidly. Yet 1 tonne of cement production is producing almost 1 tonne of carbon dioxide (CO_2). This gas causes global warming as well as environmental pollution. So, to save our environment, finding the alternative to cement is the main focus. Also due to industrialization, most of the industrial wastes (fly ash, ground granulated blast furnace slag, etc.) are being generated on a large scale. Waste disposal is the main problem. Waste utilization and production of an alternative to concrete is the main aim of the researchers. Fly ash was used for various purposes such as cement manufacturing, fly ash bricks, etc. But there is a huge quantity of fly ash generated from the industries. Geopolymer is an innovative research invention which is the alternative to cement (Palomo et al., 1999; Van Deventer et al., 2007). The source materials that are rich in silica and alumina convert into a binder using alkaline activators such as sodium hydroxide or potassium hydroxide combined with sodium silicate or potassium silicate . Geopolymer concrete is the preferred type to cement concrete.

It reduces the demand for cement concrete. Ultimately, it can decrease the greenhouse gas emissions. As it uses waste materials, it can also solve the disposal problem. This fly ash is utilized for the coarse aggregate by making it into pellets. This pellet is known as sintered fly ash aggregate or lightweight aggregate or low-density aggregate. The performance of lightweight aggregate and its effect on strength was studied in previous literature (Swamy et al., 1981; Wasserman et al., 1997; Zhang et al., 1990).

The influence of water to binder and naphthalene sulphonate polymer-based superplasticizers based on GPC was also studied in previous literature. A 14M (molar) NaOH solution is mixed with Na_2SiO_3 solution with 0.66 for the alkaline liquid (Albitar et al., 2015). The GPC focused on low-calcium fly ash concrete and the binder content of the specimens ranged between 450–700kg/m^3 and cured the specimen cubes under ambient conditions (Nath et al., 2015). Concrete with fiber-reinforced light aggregate shows the better mechanical strength than that of Ordinary Portland Cement (OPC) concrete (Kayali et al., 2003). Lightweight GPC from a lightweight block has been proven to be a good construction material for partition walls (Sata et al., 2012).

This paper explores the impact of Low-Density Aggregate (LDA) in fly ash GPC as a partially substituted natural coarse aggregate (NCA). With varying percentage of LDA, its workability, density, compressive strength, split tensile strength and flexural strength were studied.

20.2 Materials and Method

20.2.1 Fly Ash

Fly ash used in this experimental study was obtained from Hindalco Industries Ltd, Hirakud, and it satisfies the specification requirement as per IS 3812: 2013. The resulting values of physical properties like specific gravity, according to IS 1727: 1967, class and color of fly ash are listed in Table 20.1. Figures 20.1 and 20.2 represent the visual image and microscopic image collected from SEM analysis, respectively. The EDX analysis report regarding the elemental composition is shown in Table 20.2.

The size of fly ash elements varies depending on the sources of its formation. Some ashes may be finer or coarser compared to OPC particles. From Table 20.2 it was detected that the calcium content of the fly ash was low, that is 0.37%, and a considerable amount of alumina and silica constituting 10.39% and 17.62% of the weight of the whole sample, respectively was found, satisfying the criteria of source material for GPC.

TABLE 20.1

Physical Properties of Fly Ash

Property	Observation
Specific gravity	2.47
Color	Dark gray
Class	F

FIGURE 20.1 Shows the image of fly ash.

FIGURE 20.2 Shows the SEM image of fly ash.

TABLE 20.2

Elemental Composition of Fly Ash

Element	Weight %	Atomic %
C	14.71	21.78
O	52.68	58.56
Al	10.39	6.85
Si	17.62	11.16
K	0.59	0.27
Ca	0.37	0.16
Ti	1.08	0.40
Fe	2.57	0.82

20.2.2 Fine Aggregates

The observed values of percentage of mass retained in each sieve after the sieve analysis of the fine aggregate are satisfying the zone-II according to IS 2386(Part-1): 1963a.

20.2.3 Coarse Aggregate

The value of coarse aggregate obtained from the physical property tests by referring IS 2386 (Part-3): 1980 listed in Table 20.3.

20.2.4 Low-Density Aggregate

The sintered fly ash low-density aggregates (LDA) obtained for the investigation were from Indian Mineralogy and Ferro Alloy (IMFA) private Ltd. Choudwar, Cuttack. The LDA should be pre-saturated

TABLE 20.3

Physical Properties of Coarse Aggregate

Properties	Value
Specific gravity	2.85
Water absorption	0.7%

TABLE 20.4

Physical Properties of Coarse Aggregate

Properties	Value
Shape	Round
Bulk density	800kg/m^3
Specific gravity	1.57
Water absorption	14%

FIGURE 20.3 Shows the image of LDA.

FIGURE 20.4 Shows the SEM image of LDA.

with water. Hence, the aggregates immersed in water for 30–35 minutes before the casting and used after 5 minutes of drying at room temperature. It complies with all the specifications mentioned on ASTM C330-99 and IS 9142-1979. The physical properties of low-density aggregate such as bulk density, specific gravity, water absorption, obtained by the experiment are listed in Table 20.4. Figures 20.3 and 20.4 show the visual image and SEM image of LDA. Table 20.5 represents the elemental composition of the LDA powder from EDX analysis.

TABLE 20.5

Elemental Composition of LDA

Element	Weight %	Atomic %
C	9.29	15.92
O	57.84	70.41
Al	14.02	10.70
Si	22.56	16.53
K	0.83	0.43
Ca	0.88	0.45
Ti	0.76	0.33
Fe	3.11	1.15

20.3 Alkaline Solutions

Throughout the geopolymerization process, alkaline solution plays a crucial role as it dissolves the silicon and aluminum minerals from the source material to form the monomers that further combine to form a strong polymer chain. For geopolymer synthesis, a mixture of sodium hydroxide (NaOH) and sodium silicate (Na_2SiO_3) solutions are used as alkaline liquid. NaOH is a colorless crystalline solid with 98% purity of high alkaline nature. These are available in the form of pellets which are dissolved in distilled water to form the solution by a proportion depending upon the required concentration of the solution. Sodium silicate (Na_2SiO_3) solution (SiO_2 = 29.42%, Na_2O = 14.27%, Water = 56.32% and SiO_2/Na_2O ratio = 2.06) was collected from chemical supplier, Raipur, Chattishgarh.

20.4 Mixing, Casting and Curing

All the GPC mixes were prepared by replacing NCA with LDA at 5%, 10%, 15% and 20%. The casting of concrete cubes was done at room temperature. The casting was done by taking a14M NaOH concentrated solution with the ratio of $NaOH/Na_2SiO_3$ as 0.4. The 14M NaOH solution was prepared by taking 560gm (14 × 40) of solid NaOH pellets with 98% purity and mixed with Na_2SiO_3 solution before 24 hours to the casting as heat is produced during the chemical reaction of NaOH pellets with water. All the dry ingredients, i.e., source material (fly ash), NCA, LDA, and fine aggregates were weighted and mixed as per the mix proportion given in Table 20.6. Then alkaline liquid and extra water (required for workability) were poured in the mixer until the uniform mix was found. Upon preparation, for compressive, split tensile and flexural strength respectively, the cubes were molded in size 100mm × 100mm × 100mm and cylinders in size 150mm × 300mm or prism size 100mm × 100mm × 500mm. The samples were demolded and set at a temperature of 70°C for 24 hours in the oven for curing.

TABLE 20.6

Mix Proportion of GPC Mixes in (kg/m^3)

GPC Mix	FA	Sand	NCA	LDA	Alkaline	NaOH	Na_2SiO_3	Water	Solution/ binder ratio
GPL0	410	590	1100	-	205	147	58	70	0.5
GPL5	410	590	1045	55	205	147	58	70	0.5
GPL10	410	590	990	110	205	147	58	70	0.5
GPL15	410	590	935	165	205	147	58	70	0.5
GPL20	410	590	880	220	205	147	58	70	0.5

20.5 Result and Discussion

20.5.1 Workability

The workability of a fresh GPC mix was tested using the slump cone approach as described in IS 1199, 1959 and the results are provided in Table 20.7. GPC mixes show very low slump value because of the addition of sodium silicate gel which is sticky and highly viscous. Extra water of 70kg/m³ of concrete have thus been applied to increase the workability (Jena et al., 2019).

From the above table, it can be observed the sticky behavior of the GPC, which recorded a low to moderate slump value for all types of mixes. But the slump value increases with an increase in the LDA proportion. This behavior may be justified because of the round shape of the lightweight aggregates and the use of saturated aggregates before the preparation of the mixture.

20.5.2 Density (ρ)

In Table 20.8, the fresh and hardened densities of all GPC blends are shown. It was found that the density of the hardened concrete was less than that of the fresh concrete due to the heat curing of the test, which caused the water to evaporate in the fresh geopolymer mixture (Jena et al., 2019). Therefore, with the rise in the percentage of LDA replacement, the density of both fresh and hard concrete was found to decrease, which is an advantage of the low-density aggregate.

20.5.3 Compressive Strength (CS)

Figure 20.5 displays the variability in CS with different LDA percentages. It was found that the CS of concrete for 5% and 10% replacement is marginally increased, but further rises in the LDA percentage subsequently decreases in strength as in 15% and 20%. Throughout the breakdown of the samples by means of the Compressive Testing Machine, it was found that in the event of a rise in the percentage of the fly ash aggregate, the disintegration of the specimen was more than 5% and 10% of the samples suggesting a concrete matrix or matrix aggregate bond failure to substitute a higher percentage.

This can be described as the increase in the fly ash content in the mix (using fly ash LDA) as a result of less availability of sufficient alkali solution for reaction due to which the concrete failure or bond failure

TABLE 20.7

Slump of GPC Mixes

Design mix	% of LDA replacement	Slump (in mm)
GPL0	00	63
GPL5	05	68
GPL10	10	75
GPL15	15	93
GPL20	20	99

TABLE 20.8

Fresh and Dry Densities (kg/m³) of GPC Mixes

Design mix	Fresh density	Hardened density
GPL0	2,530	2,438
GPL5	2,445	2,340
GPL10	2,398	2,303
GPL15	2,352	2,287
GPL20	2,307	2,217

Experimental Investigation on Geopolymer Concrete 171

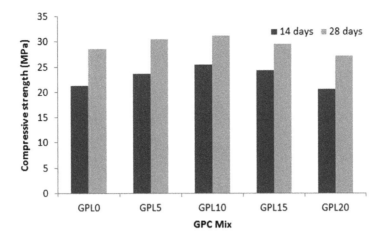

FIGURE 20.5 Variation in compressive strength with varying percentage of LDA.

was found. The decrease in the strength can also be due to the use of round shaped aggregates instead of natural flaky and elongated aggregates (Kayali et al., 2003).

20.5.4 Split Tensile Strength

The findings of the split tensile strength test being carried out are shown in Figure 20.6. The GPC shows low split tensile strength due to porous LDA (Kockal & Ozturan, 2010). From the figure, it was observed that the decrease in tensile strength with an increase in the LDA percentage above 10%. But the strength was increased with 5% and 10% replacement showing good results.

20.5.5 Flexural Strength

The flexural strength test outcomes of all GPC mixtures with different LDA percentages are shown in the Figure 20.7. Flexural strength of geopolymer mix at 10% LDA replacement was found to be of maximum strength than that of the control concrete. Flexural strength was increased as the replacement goes on increasing up to 10%, then a sudden fall was noticed. But the strength of GPL15 was more than the control concrete. This could be due to the increased amount of fly ash quantity in the GPC due to the addition of sintered fly ash aggregates.

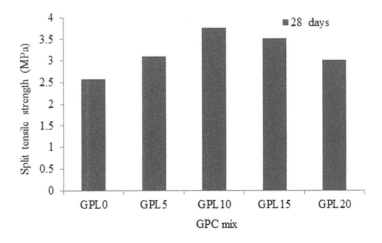

FIGURE 20.6 Variation in split tensile strength with varying percentage of LDA.

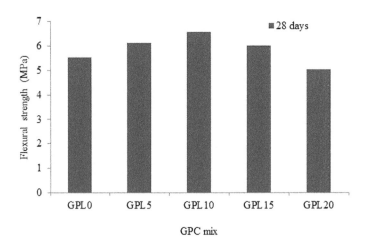

FIGURE 20.7 Variation in flexural strength with varying percentage of LDA.

20.6 Conclusion

This paper provides a thorough analysis of the conclusions that can be drawn from the current research on low-calcium fly ash-based geopolymer concrete with lightweight aggregates that substitute natural aggregates at 5%, 10%, 15% and 20%. The conclusions obtained from these discussions are explained below:

- The workability of GPC was always laid in low to medium range because of Na_2SiO_3 gel and its sticky behavior. But the incorporation of fly ash aggregates cause a slight increase in strength due to their round shapes, which are incapable of contributing any type of resistance by interlocking. The use of saturated LDA can also be the reason as it possesses a high value of water absorption due to porous.
- A lower value of fresh, as well as hardened density, was found in the LDA used specimens because of the lower specific gravity of these aggregates from the natural aggregates. Dry density, unlike every concrete mix, was found lower than the fresh density due to the evaporation water from the polymer matrix during the curing period.
- Compressive strength of LDA specimens was found showing an increased result of up to 10% but later it falls with an increased percentage of replacement.
- The split tensile strength and flexural strength were found to increase with the increase in the percentage of LDA due to the porous nature of light aggregate.

REFERENCES

Albitar, M., Visintin, P., Ali, M.S.M., and Drechsler, M., 2015. Assessing behaviour of fresh and hardened geopolymer concrete mixed with class-F fly ash. *KSCE Journal of Civil Engineering*, 19(5), 1445–1455.

Ganesan, N., Abraham, R., Raj, S.D., and Sasi, D., 2014. Stress–strain behaviour of confined geopolymer concrete. *Construction and Building Materials*, 73, 326–331.

IS: 2386, 1963a. *Indian Standard Specification, Methods of Test for Aggregates for Concrete: Part 1, Particle Size and Shape*, Bureau of Indian Standards, New Delhi.

IS: 2386, 1963b. *Indian Standard Specification, Methods of Test for Aggregates for Concrete: Part 3, Specific Gravity, Density, Voids, Absorption and Bulking*, Bureau of Indian Standards, New Delhi. [Reaffirmed in 2002].

IS - 1199 1959. *Methods of Sampling and Analysis of Concrete*, Bureau of Indian standards, New Delhi, India.

Jena, S., Panigrahi, R., and Sahu, P., 2019. Effect of silica fume on the properties of fly ash geopolymer concrete. In *Sustainable Construction and Building Materials* (pp. 145–153). Springer, Singapore.

Kayali, O., Haque, M. N., and Zhu, B., 2003. Some characteristics of high strength fiber reinforced lightweight aggregate concrete. *Cement and Concrete Composites*, 25(2), 207–213.

Kockal, N. U., and Ozturan, T., 2010. Effects of lightweight fly ash aggregate properties on the behavior of lightweight concretes. *Journal of hazardous materials*, 179(1–3), 954–965.

Nath, P., Sarkar, P.K., and Rangan, V.B., 2015. Early age properties of low-calcium fly ash geopolymer concrete suitable for ambient curing. *Procedia Engineering*, 125, 601–607.

Palomo, A., Grutzeck, M.W., and Blanco, M.T., 1999. Alkali-activated fly ashes: Cement for the future. *Cement and Concrete Research*, 29(8), 1323–1329.

Sata, V., Sathonsaowaphak, A., and Chindaprasirt, P., 2012. Resistance of lignite bottom ash geopolymer mortar to sulfate and sulfuric acid attack. *Cement and Concrete Composites*, 34(5), 700–708.

Swamy, R.N., and Lambert, G.H., 1981. The microstructure of lytag aggregate. *The International Journal of Cement Composites and Lightweight Concrete*, 3(4), 273–282.

Van Deventer, J.S.J., Provis, J.L., Duxson, P., and Lukey, G.C., 2007. Reaction mechanisms in the geopolymeric conversion of inorganic waste to useful products. *Journal of Hazardous Materials*, 139(3), 506–513.

Wasserman, R., and Benthur, A., 1997. Effect of lightweight fly ash aggregate microstructure on the strength of concrete. *Cement and Concrete Research* 27(4), 525–537.

Zhang, M.H., and Gjorv, O.E., 1990. Characteristics of lightweight aggregates for high-strength concrete. *ACI Materials Journal*, 88(2), 150–157.

21

Strength Development in Ferrochrome Ash-Based Geopolymer Concrete

Jyotirmoy Mishra
Veer Surendra Sai University of Technology, India

S. K. Das
CSIR- Institute of Minerals and Materials Technology, India

Bharadwaj Nanda and Sanjaya Kumar Patro
Veer Surendra Sai University of Technology, India

Syed M. Mustakim
CSIR- Institute of Minerals and Materials Technology, India

CONTENTS

21.1 Introduction 175
21.2 Materials and Methods 176
    21.2.1 Materials 176
    21.2.2 Methods of Manufacturing Geopolymer Concrete 177
21.3 Results and Discussions 178
    21.3.1 Workability 179
21.4 Compressive Strength 179
    21.4.1 Effect of Source Materials and Extra Water 179
    21.4.2 Effect of Curing Temperature on Strength Development 180
    21.4.3 Efflorescence 181
21.5 Conclusions 182
Acknowledgment 182
References 182

21.1 Introduction

Emerging awareness on the generation of the industrial solid wastes like fly ash (FA), ferrochrome ash (FCA), ground granulated blast finance slag (GGBFS), and rice husk ash (RHA), etc., and growing concerns of the environmental after-effects of industrial waste disposal have encouraged the scientific community to explore new avenues for their use in the construction sectors.

    The infrastructure growth rate is directly reflected by the demand for the cement. However, cement production is an energy-intensive process which generates a massive amount of carbon dioxide gas (CO_2) that poses a threat to our environment. One tonne of cement immediately releases about 0.55 tonnes of CO_2 during its production. Further, the combustion of fossil fuels during cement production emits about 0.40 tonnes of CO_2 into the environment. Collectively, the production of one tonne of cement releases about one tonne of carbon dioxide gas to the atmosphere (Davidovits, 1994). Hence, there is an urgent need for the technology and strategy for a sustainable industrial solid waste management to use maximum waste in the construction sector for the production of low-carbon, eco-friendly concrete. The concept of

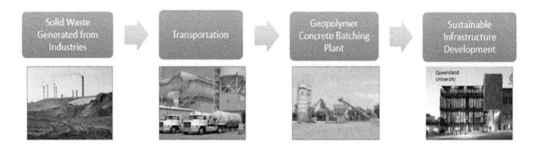

FIGURE 21.1 A geopolymer concrete based model for industrial solid waste management.

geopolymer concrete is a comparatively recent development. It is produced from the alkali-activation of various industrial waste products such as FA, GGBFS, etc. (Davidovits, 2008). Therefore, it has the potentiality to be a suitable model for the large-scale use of these aluminosilicate-rich industrial wastes as shown in Figure 21.1.

Since its introduction, several efforts have been made worldwide to synthesize geopolymers from various industrial wastes like FA, GGBFS, FCA, RHA, etc., and the results are found to be quite promising with improved mechanical strength and durability properties, and reaction mechanisms (Mishra et al. 2020; Mishra et al., 2019a, b ; Das et al., 2018; Alomayri et al., 2014).

Ferrochrome ash (FCA) is the dust collected from the gas cleaning plants of the ferrochrome industry and generally discarded as waste whose disposal is a prime concern for the industry. It is estimated that the production of ferrochrome is around 9.5 million tonnes worldwide and is also increasing at a rate of 3% annually. Production of one tonne of ferrochrome ash produces about 20–30kg of ferrochrome ash (Acharya & Patro, 2015). Few studies are made to examine the effect of FCA as partial replacement to cement in preparation of standard concrete (Acharya & Patro, 2015, 2016a, 2016b, 2016c).

The partial replacement of Portland cement by 40% FCA, and 7% lime, simultaneously resulted in the improvement of the properties of concrete (Acharya & Patro, 2016a). In another study, Acharya & Patro (2016b), found that with simultaneous usage of FCA and lime as partial replacement of cement, and ferrochrome slag as partial replacement of coarse aggregate resulted in further improvement in strength, and sorptivity of conventional concrete. Moreover, using the combination of FCA and lime significantly improved the acid resistance of the concrete (Acharya & Patro, 2016c).

On the other hand, various mineralogical studies suggest that the ferrochrome ash contains a considerable amount of aluminum, silicon, magnesium, etc., that are required for geopolymerization. Yet, the suitability of FCA has not been exploited for the preparation of geopolymer concrete, to date.

In this paper, a study is made on the use of FCA as primary source material in combination with secondary source materials like GGBFS, and fly ash to prepare geopolymer concrete. The role of FCA on the workability and compressive strength under ambient curing conditions is analyzed. The alkaline solution used for the activation of source materials consisted of 12M (mole) sodium hydroxide solution and sodium silicate solution. The effects of curing temperature, mixing of extra water, and efflorescence on strength characteristics of resulting geopolymer concrete are also discussed.

21.2 Materials and Methods

21.2.1 Materials

In this experimental work, the ferrochrome ash was employed as the primary source material in the preparation of geopolymer concrete. This was supplemented by FA, and GGBFS as the secondary source materials. The source materials viz. FCA, FA, and GGBFS were collected from the Balasore Alloys Pvt. Ltd., Balasore, Talcher Super Thermal Power Station (NTPC), Talcher, and Neelachal Ispat Nigam Ltd., Jajpur, respectively. The commercial-grade sodium hydroxide and sodium silicate were procured from

FIGURE 21.2 (a,b,c, and d) Raw materials used for the experimental investigation.

local markets. Figure 21.2 shows the images of the source materials and the alkaline solution used in the present study.

The chemical compositions of the source materials presented in Table 21.1 are determined from the X-ray fluorescence (XRF) analysis. Ferrochrome ash has a partially spherical and loose microstructure while fly ash has a highly spherical and porous structure. In comparison, GGBFS has a highly irregular and compact microstructure. The nature-originated fine and coarse aggregates conforming to Indian standards (IS 383:1970) were taken in the surface dry condition for concrete preparation.

21.2.2 Methods of Manufacturing Geopolymer Concrete

The objective of this research is to investigate the possibility of geopolymerization using ferrochrome ash as the primary source material. FA and GGBFS are added to supplement the FCA as it has low alumina and silica content. Hence, the study consisted of two phases. First, GGBFS was incorporated as supplementary source material to FCA for the preparation of geopolymer concrete. For this, the collected blast furnace slag was ground with the help of a ball mill to achieve the required fineness suitable for

TABLE 21.1

Chemical Compositions of the Source Materials

Materials	SiO$_2$	Al$_2$O$_3$	CaO	MgO	Fe$_2$O$_3$	K$_2$O	Na$_2$O	P$_2$O$_5$	TiO$_2$	Cr$_2$O$_3$
FCA	19.1	10.91	3.14	23.59	7.83	11.42	2.45	0.07	N/A	9.89
Fly Ash	60.34	30.83	0.81	0.54	3.34	1.26	0.08	0.52	1.87	0.02
GGBFS	36.28	20.38	24.1	8.07	6.64	1.02	0.38	0.05	0.72	0.07

TABLE 21.2

FCA Based Geopolymer Concrete Mix Proportions Designed in this Study

Mix Designation	FCA in kg/m$_3$	GGBFS in kg/m$_3$	Fly Ash in kg/m$_3$	Fine Aggregate in kg/m$_3$	Coarse Aggregate in kg/m$_3$	Sodium Hydroxide in kg/m$_3$	Sodium Silicate in kg/m$_3$	Extra water in liters/m$_3$
S1	394.4	98.6	-	493	1479	73.08	154.07	-
S2	295.8	197.2	-	493	1479	73.08	154.07	-
S3	394.4	98.6	-	493	1479	73.08	154.07	14.81
S4	295.8	197.2	-	493	1479	73.08	154.07	14.81
S5	295.8	98.6	98.6	493	1479	73.08	154.07	14.81

geopolymerization. In the next phase, a part of GGBFS was replaced by fly ash. The alkaline solution composed of sodium hydroxide (NaOH) and sodium silicate (Na$_2$SiO$_3$) is mixed to formulate the solution 24 hours before mixing with the source materials. Sodium hydroxide was used in the form of flakes with 98% purity. Similarly, the sodium silicate solution had a chemical composition of 32.15% SiO$_2$, 15.85% Na$_2$O, and 52% water. Literature suggests that concentration of NaOH in the range of 8M-16M have been investigated for preparation of geopolymer concrete. An optimum concentration of 12M is recommended for strength development in case of geopolymer concrete. Therefore, the concentration of sodium hydroxide used for this present study was taken to be 12M. Likewise, some preliminary investigations were conducted before this study that followed a trial-and-error method to find a suitable ratio of NaOH to Na$_2$SiO$_3$ which could facilitate better strength development in geopolymer concrete. A ratio of 1:2.1 (NaOH: Na$_2$SiO$_3$) by weight was found to provide better strength gain. Hence, in this study, a ratio of 1:2.1 (NaOH: Na$_2$SiO$_3$) by weight was used. One liter of water was mixed with solution 480gms of NaOH flakes to prepare a solution of 12M concentration (1.48kg). Thereafter, it was combined with two liters of Na$_2$SiO$_3$ solution (3.1kg) referring to the weight ratio (1:2.1). Total NaOH and Na$_2$SiO$_3$ required for 1m^3 of geopolymer concrete is 73.08kg/m^3 and 154.07kg/m^3, respectively, as shown in Table 21.2.

In total, five different mix proportions, shown in Table 21.2, were considered for the present study. Six cube specimens were prepared for each mix designation. The source materials were first dry-mixed with the inert aggregates and then, the formulated alkali activator and extra water were added gradually while the mixing continued. The total procedure lasted for about 11 minutes. The workability of the fresh geopolymer concrete mix is carried out by standard slump cone test as per the provisions of IS 1199:1959. Then, the fresh geopolymer concrete mixes were poured into 150mm cube molds as specified by the Indian standards (IS 10086:1982, 1982) and left to be cured under ambient temperature conditions inside the laboratory (temperature: 23 ± 3°C, and relative humidity: 60%). Assessment of compressive strength is made using a 2,000kN capacity compression testing machine as per the guidelines of IS 516:1959, 1959.

21.3 Results and Discussions

Assessment of the workability and compressive strength are made for the different geopolymer mixes. Further, the effect of curing temperature and influence of efflorescence on strength properties of the geopolymer mixture are also studied for these geopolymer mixes.

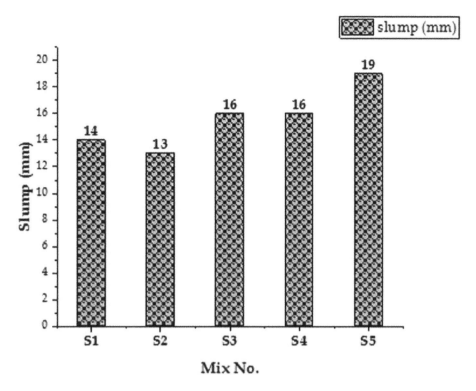

FIGURE 21.3 Slump values of the geopolymer concrete mixes.

21.3.1 Workability

The slump values achieved for all the five GPC mixtures are represented in Figure 21.3, wherein it may be seen that the slump value ranges between 14mm to 19 mm, corresponding to a very stiff mix. However, somewhat improvement in the slump value is observed due to the addition of the fly ash. The spherical particle size of the fly ash acted like ball bearings to improve the workability. Similar to the findings of other research (Wardhono et al., 2015), it may also be seen from the figure that the addition of extra water has slightly improved the workability in a very small quantity. Further, it is observed that the additional fly ash in the system made the geopolymer matrix stronger and provided better polymeric gel formation.

21.4 Compressive Strength

21.4.1 Effect of Source Materials and Extra Water

The cube specimens cast with geopolymer concrete of different mix designations were tested for the analysis of compressive strength development. Generally, geopolymer concrete cubes are recommended for elevated temperature curing. However, achieving a continuous elevated temperature may not be feasible for *in situ* constructions and, therein, the ambient curing may be the only option. For enabling strength development in ambient curing conditions, the addition of GGBFS is often recommended. Accordingly, in the present study, GGBFS was added at 20% and 40% replacement levels of FCA. As the workability of geopolymer concrete was lower, an attempt was made to add extra water during mixing, and its effect on compressive strength was observed.

The 7-, 14-, and 28-day compressive strength of different FCA-based GPC mixes are illustrated in Figure 21.4. It is observed that; the highest compressive strength was achieved by mix S5 that contains 20% of fly ash along with 20% GGBFS in FCA-based geopolymer concrete. This suggests that the addition of fly ash into the system improved the strength development of FCA-based geopolymer cubes.

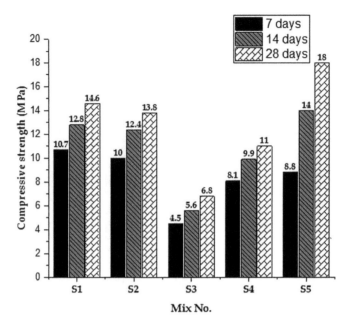

FIGURE 21.4 Compressive strengths of different geopolymer mixes.

It is also observed from the figure that the strength improvement of the S5 geopolymer mix from 7 to 28 days is 104.54%. In comparison, the strength improvement between these two ages is 36.4%, 38.0%, 51.1%, and 35.0%, respectively, for the mixes containing GGBFS, and FCA, namely S1, S2, S3, and S4 designations.

It is observed from the past research that, GGBFS-based geopolymer concrete develops strength in ambient temperature due to the generation of calcium silicate hydrate (C-S-H) gel which is one of the principal hydration products (Song et al., 2000). This is further reinforced by the findings of Nath & Sarker (2014), which suggests that the strength development of geopolymer concrete is mainly because of the synergetic effect of calcium aluminosilicate hydrate (C-A-S-H), calcium silicate hydrate (C-S-H) and sodium aluminosilicate hydrate (N-A-S-H) gels.

On the contrary, several studies suggest the FA-based geopolymer concrete exclusively gains strength at an elevated temperature (Puertas et al., 2000; Lee & van Deventer, 2002; Wallah & Rangan, 2006). Hence, the early strength development (7 days) for the mix S5 containing fly ash is found to be lower than the mixes S1, and S2, although the 28-day strength is found to be better. However, the addition of fly ash increases the aluminosilicate content and balances the lime, magnesia quantity in the resulting FCA-GGBFS-FA source mix making it suitable for geopolymerization.

The effect of the extra water content has also been investigated in the present research. The first two mixes (S1, and S2) were prepared without any additional source of water. Only the water present in the alkali activator is used for geopolymer reaction and strength development for these two mixes. As the resulting geopolymer mix was found to be very stiff without extra water, extra water at the rate of 14.81 liters/m^3 of geopolymer concrete was added while mixing the ingredients (specimens S3, S4, and S5).

It is evident from the figure that, the inclusion of extra water decreased the compressive strength of the resulting geopolymer mixes. The reductions in compressive strength at the age of 28 days due to the extra water are found to be 53.4%, and 20.3% between S1 and S3, and S2 and S4, respectively. This is in the line of earlier findings (Hadi et al., 2017; Wardhono et al., 2015; Mishra et al., 2019a, b).

21.4.2 Effect of Curing Temperature on Strength Development

The curing temperature is an important factor that regulates the strength development and other mechanical properties of the geopolymer concrete. The elevated curing temperatures facilitate the rapid polymerization process that attributes to the early strength development (Rangan 2014). A range of curing temperatures of 40°C to 140°C, and a curing duration about 10 hours to 72 hours has been suggested in the past. However, the heat-curing method may be suitable only for pre-cast concrete applications. For *in situ* cast applications, it is impracticable to subject a structure to the constant temperature for a prolonged period. Therefore, to lower the curing temperature, a calcium-rich material, i.e., GGBFS in the present case, is used as supplementary source material to the FCA.

In the present study, the cast cube specimens are subjected to ambient curing conditions at room temperature (23 ± 3°C) and relative humidity approximately 60%. It is noticed that the strength developments in the geopolymer concrete specimens were slow but steady. The addition of GGBFS accelerated the rate of formation of C-S-H, C-A-S-H, and N-A-S-H gel that predominately contributes to the strength. Although, it is observed that the 28-day strength could not improve much. Therefore, further research is required to achieve a high strength geopolymer concrete under ambient curing conditions.

21.4.3 Efflorescence

Efflorescence refers to a process in which the material exhibits some salts/complex compounds on its surface due to chemical reactions (carbonation/ leaching). It is a very common problem observed in both cement concrete and geopolymer concrete. The geopolymer concrete undergoes efflorescence under high-humidity atmospheric conditions. Škvára et al. (2009) suggested that the weak bond of Na+ inside the aluminosilicate polymers structure causes the efflorescence in geopolymer concrete. The efflorescence product mainly consisted of several sodium hydrate compounds, like $Na_2CO_3.nH_2O$, $NaHCO_3.nH_2O$, $Na_6(SO_4)(CO_3, SO_4).nH_2O$.

In present experimental work, FCA-based geopolymer cube samples are found to be affected by efflorescence after the 28-day curing period at low temperature and high humidity of 60%. Efflorescence, the whitish colored compounds, was formed in the surface of the geopolymer concrete cube open to the atmosphere as displayed in Figure 21.5. It indicates the presence of carbonation on the surface of the geopolymer cubes.

For a better understanding of the efflorescence product, some samples were collected from the cube surface subjected to the efflorescence. Then XRF analysis was performed on these samples to identify the efflorescence compounds, shown in Table 21.3, wherein it may be figured that the major part of the efflorescence sample is sodium oxide (Na_2O). This indicates that there was a poor geopolymerization of the precursor materials which may be due to high humid and low-temperature conditions. The poor geopolymerization has affected the strength development of the specimens at later ages. Therefore, the efflorescence negatively affects both the strength and microstructure of the geopolymer concrete. The control of efflorescence in geopolymer concrete acts as a major challenge to the researchers. Few studies (Longhi et al., 2019) suggest the inclusion of additives like magnesia (MgO) to control efflorescence. Yet, extensive and specific research is required in this aspect to find a suitable remedy for this.

FIGURE 21.5 Efflorescence in geopolymer concrete cube specimens.

TABLE 21.3
Compounds Identified in Efflorescence Product

Chemical Compounds	Na$_2$O	Al$_2$O$_3$	SiO$_2$	MgO	SO$_3$	Cr$_2$O$_3$	Cl	CaO	K$_2$O	Fe$_2$O$_3$	Ti$_2$O	K$_2$O
Weight (%)	52.59	3.87	13.40	1.89	3.49	1.80	10.00	2.25	5.05	4.32	0.49	5.05

21.5 Conclusions

This paper discusses a possible industrial solid waste management strategy for the ash generated in the ferrochrome industry. The outcomes of this study on the low-carbon geopolymer concrete using the ferrochrome ash supplemented by GGBFS and FA may be summarized as follow:

- FCA can be employed as the primary source material for manufacturing of geopolymer concrete under ambient room temperature conditions. The addition of fly ash has significantly improved the compressive strength of the FCA-GGBFS-based geopolymer concrete mix.
- The fly ash provides additional silica and alumina, which increased the volume of aluminosilicate compounds in the blended source materials, and enhanced their dissolution in the alkaline solution. This, in turn, increased the N-A-S-H, C-A-S-H, and C-S-H gel formation in FCA-GGBFS-FA-based geopolymer concrete due to which the strength improved. Besides, the addition of fly ash also enhanced the workability of the fresh geopolymer concrete mix.
- Additional water did not improve workability by much. However, it negatively impacted the strength development of the resulting geopolymer concrete mix. Further, discussions on curing regime and efflorescence on compressive strength are also made in this paper.
- The mineralogical analysis of efflorescence products revealed an increased presence of sodium oxide in the efflorescence product. Further, the cause of efflorescence and a few measures to control the efflorescence is also discussed herein.

It is possible to valorize these industrial wastes by their effective utilization in the field of the construction industry. The results found in the present study may serve the first step in that direction. However, before valorization of these industrial wastes, we need to study further in the direction of durability and other mechanical properties before establishing FCA as effective source material for geopolymerization.

ACKNOWLEDGMENTS

The authors are grateful to Mr. R.S. Krishna and other technical staffs of CSIR-IMMT, Bhubaneswar for their assistance during the experiments. The help of Dr. R. Mohapatra, Mr. S. Kundu of Balasore Alloys Ltd., Balasore, and Mr. S. Sathpathy, of Neelachal Ispat Nigam Ltd., Jajpur in arranging the source materials are also acknowledged.

REFERENCES

Acharya, P.K., and Patro, S.K. (2015), 'Effect of lime and ferrochrome ash (FA) as partial replacement of cement on strength, ultrasonic pulse velocity and permeability of concrete', *Construction and Building Materials*, 94:448–457.
Acharya, P.K., and Patro, S.K., (2016a), 'Sustainable use of ferrochrome ash (FCA) and lime dust in concrete preparation', *Journal of Cleaner production*, 131:237–246.
Acharya, P.K., and Patro, S.K. (2016b), 'Utilization of ferrochrome wastes such as ferrochrome ash and ferrochrome slag in concrete manufacturing', *Waste Management Research*, 34:764–774.
Acharya, P.K., and Patro, S.K., (2016c), 'Acid resistance, sulphate resistance and strength properties of concrete containing ferrochrome ash (FA) and lime', *Construction and Building Materials*, 120:241–250.

Alomayri, T., Shaikh, F.U.A., & Low, I.M. (2014), 'Synthesis and mechanical properties of cotton fabric reinforced geopolymer composites', *Composites Part B: Engineering*, 60, 36–42.

Das, S.K., Mishra, J. & Mustakim, S.M. (2018), 'Rice husk ash as a potential source material for geopolymer concrete: A review', *International Journal of Applied Engineering Research*, 13(7): 81–84.

Davidovits, J. (1994), 'Global Warming Impact on the Cement and Aggregates Industries', *World Resource Review*, 6(2): 263–278.

Davidovits, J. (2008) *Geopolymer Chemistry and Application*. 2nd edn. Saint-QuentinFrance: Geopolymer Institute.

Hadi, M.N.S., Farhan, N.A., and Sheikh, M.N. (2017), 'Design of geopolymer concrete with GGBFS at ambient curing condition using Taguchi method', *Construction and Building Materials*, 140: 424–431.

IS 10086:1982 (1982), 'Specification for moulds for use in tests of cement and concrete', Bureau of Indian Standards, New Delhi.

IS 1199:1959 (1959), 'Methods of sampling and analysis of concrete', Bureau of Indian Standards, New Delhi.

IS 383:1970 (1970), 'Specification for coarse and fine aggregates from the natural sources for concrete', Bureau of Indian Standards, New Delhi.

IS 516:1959 (1959), 'Method of tests for strength of concrete', Bureau of Indian Standards, New Delhi.

Lee W.K.W., van Deventer J.S.J. (2002), 'The effect of ionic contaminants on the early-age 760 properties of alkali-activated fly ash-based cements', *Cement and Concrete Research*, 32 (4): 577–761

Longhi, M.A., Zhang, Z., Rodríguez, E.D., Kirchheim, A.P. & Wang, H, (2019), 'Efflorescence of alkali-activated cements (geopolymers) and the impacts on material structures: A critical analysis', *Frontiers in Materials*, 6:89.

Mishra, J., Das S. K., Mustakim S.M., (2019b), 'Geopolymer Technology for the Promotion of Sustainable Built Environment in Future India', In: *Advances in Civil Engineering*, vol-3, AkiNik Publications, Delhi, India.

Mishra, J., Das S.K., Singh, S.K., & Mustakim, S.M. (2019a), 'Development of geopolymer concrete for the protection of environment: A greener alternative to cement', *SSRG International Journal of Civil Engineering*, 6(3): 41–47.

Mishra, J., Das, S.K., Krishna, R.S., Nanda, B., Patro, S.K., Mustakim, S.M., (2020), 'Synthesis and characterization of a new class of geopolymer binder utilizing ferrochrome ash (FCA) for sustainable industrial waste management', *Materials today: proceedings*, 33:5001–5006.

Nath, P., and Sarker, P.K. (2014), 'Effect of GGBFS on setting, workability and early strengthproperties of fly ash geopolymer concrete cured in ambient condition', *Construction and Building Materials*, 66:163–171.

Puertas, F., Martinez-Ramirez, S., Alonso, S., & Vazquez, T. (2000), 'Alkali-activated fly ash/slag cement Strength behaviour and hydration products', *Cement and Concrete Research*, 30:1625–1632.

Rangan, B.V. (2014), 'Geopolymer concrete for environmental protection', *The Indian Concrete Journal*, 88(4):41–48.

Škvára, F., Kopecký, L., Mysková, L., Smilauer, V., Alberovská, L., and Vinsová, L. (2009), 'Aluminosilicate polymers - influence of elevated temperatures, efflorescence', *Ceramics Silikaty*, 53(4):276–282.

Song, S., Sohn, D., Jennings, H.M., & Mason, T.O. (2000), 'Hydration of alkali-activatedground granulated blast furnace slag', *Journal of Material Science*, 35:249–257.

Wallah, S. and Rangan, B.V. (2006), 'Low-calcium fly ash-based geopolymer concrete Long-term properties', Report 2, Faculty of Engineering, Curtin University of Technology, Perth, Australia.

Wardhono, A., Law, D.W., & Strano, A. (2015), 'The strength of alkali-activated slag/fly ash mortar blends at ambient temperature', *Procedia Engineering*, 125:650–656.

22 Investigation on Strength Factor of Composite Concrete Using Quarry Dust and Artificial Aggregates

Sudheer Ponnada, Partheepan Ganesan, and G. Praveen
MVGR College of Engineering (Autonomous), Andhra Pradesh, India

CONTENTS

22.1 Introduction 185
22.2 Materials and Methods 186
      22.2.1 Quarry Dust 186
22.3 Comparison of Coarse Aggregate and Artificial Aggregate 187
22.4 Experimental Programme for the Present Study 188
22.5 Experimental Work 188
22.6 Discussions and Conclusions 189
References 190

22.1 Introduction

Generally, conventional concrete is a standardised mix of cement, sand, aggregate and water. Properties of aggregate change the durability and performance of concrete, so aggregate is an important ingredient of concrete (Sudheer et al., 2016a,b). The highest-used coarse aggregate is mineral aggregate deposits obtained from mining, which include sand, gravel and stone, and the most-utilised fine-aggregate is conventional river sand or pit-sand. Fine and coarse-aggregate comprise 75% of the total-volume of concrete (Baskaran et al., 2017, Francesco & Raffaele, 2013, Harilal & Job, 2012). Hence, it is essential to acquire and use good and the correct type of aggregate in concrete for its strength, durability and many other aspects as the aggregate forms the major part of the matrix/mortar (Jayalakshmi & Basil, 2016, Sudheer et al., 2017). The worldwide utilisation of natural fine aggregate as sand is very high. In general, the necessity of conventional aggregate is high in emerging countries to fulfil the fast-infrastructural development (Kavitha et al., 2015, Narendra & Gangha, 2015). In this position, emerging countries, like India, Brazil, South Korea, etc., are facing scarcity in excellent quality aggregates. Especially in India, natural aggregate banks are being reduced and producing serious risk to the environment (Jayalakshmi & Basil, 2016).

Increased mining and extraction of natural aggregate cause many problems such as noise, air pollution, blasting effects, sedimentation, losing water, loss of habitat, formation of sink holes, and disturbance in aquatic life as well as agriculture due to reducing the underground water table, etc., In the past decades, the over-usage of the naturally occurred aggregate has led to an increase in their cost variably and further leading to the increase in outflow of constructions (Hiraskar & Chetan, 2013). In this condition, research is initiated for economical and easily obtainable alternate materials to natural-sand and aggregate (Sudheer et al., 2018). Several other materials are used as an alternative to natural-sand such as fly-ash, slag-limestone, etc., as partial replacement, and materials used as alternatives for natural aggregate are quartz sandstone, rubber, etc., (Sudheer et al., 2020). However, shortage in good quality is the most important constraint in the above all materials (Palankar et al., 2015) (Pavan Kumar & Mahesh, 2012). Nowadays, sustainable-infrastructural development requires that the alternate-material should comply with technical requisite of aggregates and they should be available abundantly (Ravisankar et al., 2015, Shareef et al., 2015, Shareef et al., 2019).

For the present scenario, a sample investigation tried to successful utilise quarry dust as a fine aggregate and fly ash with binder as a coarse aggregate that would turn the waste that causes disposable problem into a valuable resource. The utilisation will also reduce the strain on the allocation of natural aggregate which will also lower the cost of concrete.

22.2 Materials and Methods

For present work cement (OPC 53 grade), fine-aggregate, quarry-dust, coarse-aggregate and artificial aggregate are used and physical properties for the same are listed below in Table 22.1 and Table 22.2.

22.2.1 Quarry Dust

Quarry-dust is a by-product of the coarse aggregate from crushing-process which is a focused material to utilize as aggregates for concreting objective, particularly as fine-aggregates (FA). In quarrying actions, the rock has been squashed into various-sizes; the dust generated all through the process is called quarry-dust and it is produced as waste as shown in Figure 22.1.

For present work, quarry dust was obtained from Zonnada Quarry, located near to Vizianagaram, Andhra Pradesh, India. Physical properties of quarry dust collected from above mentioned quarry was tested in the laboratory and presented in Table 22.1.

TABLE 22.1

Properties of Quarry Dust

S. No	Properties	Natural Fine aggregate	Quarry-dust	Code of practice
1.	Specific gravity	2.60	2.37	IS:2386 (Part-3)
2.	Fineness modules	Zone–II	Zone–II	IS:383
3.	Bulk density (kg/m^3)	1,380	1,690	IS:2386 (Part-3)

TABLE 22.2

Mechanical Properties of Coarse Aggregate and Artificial Aggregate

S. No	Properties	Natural coarse aggregate	Artificial coarse aggregate	Code practice
1.	Specific gravity	2.76	1.16	IS:2386 (Part-3)
2.	Bulk density (kg/m^3)	1593	925	IS:2386 (Part-4)
3.	Water Absorption (%)	1.13	0.55	IS:2386 (Part-4)
4.	Impact Value (%)	18.3	6.65	IS:2386 (Part-4)
5.	Crushing value (%)	23.4	5.58	IS:2386 (Part-4)
6.	Abrasion value (%)	32.4	8.85	IS:2386 (Part-4)

Investigation on Strength Factor 187

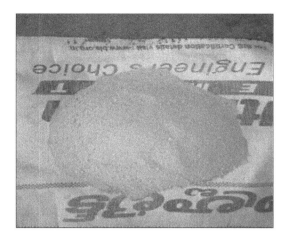

FIGURE 22.1 Quarry dust.

22.3 Comparison of Coarse Aggregate and Artificial Aggregate

For the present investigation, 20mm nominal size conventional coarse aggregates (CA) was used and the same was obtained from Zonnada Quarry. Artificial aggregates prepared by a blend of fly ash and epoxy with indigenous equipment are shown in Figure 22.2(a) and Figure 22.2(b). Physical properties of normal and artificial coarse aggregate are tested in the laboratory and compared in the Table 22.2.

FIGURE 22.2 Artificial aggregate samples with manufacturing equipment.

22.4 Experimental Programme for the Present Study

For present study, no proper code of practice is available to study about concrete mix and compressive strength for new composite concrete. For this we adopt a conventional mix design practice as per IS: 10262 for M30 grade to fix the required percentages of ingredients to lay the specimens using both conventional and non-conventional concrete. Mix design conforming to IS standard are presented in Table 22.3, and the trial mixes used for the present-investigation are presented in Table 22.4.

22.5 Experimental Work

For present work a total of seven trial mixes were tried with different material percentages as shown in Table 22.4. Average compressive strength of all mixes (i.e., three cubes for each mix) at an age of 7 days are shown in the below Table 22.5.

TABLE 22.3

Mix Design Adopted for the Present Study

Type of Material	Cement (kg/m^3)	Water (kg/m^3)	FA (kg/m^3)	CA (kg/m^3)	W/C ratio
Proportion Required	410	197	686.50	882	0.46

TABLE 22.4

Material Proportions for Present Study

Type	Cement (%)	FA (%)	CA (%)	Artificial Aggregate (%)	Quarry Dust (%)
M1	100	100	100	–	–
M2	100	100	–	100	–
M3	100	100	50	50	–
M4	100	–	100	–	100
M5	100	–	–	100	100
M6	100	50	100	–	50
M7	100	50	–	100	50

TABLE 22.5

Average Compressive Strength Values for the Trial Mixes

Type of Mix	M1	M2	M3	M4	M5	M6	M7
Avg. Compressive strengths at an age of 7 days (MPa)	23.60	20.30	22.80	23.40	23.10	23.80	22.10

22.6 Discussions and Conclusions

Following discussions are made from the test results from the above Table 22.5 and below Figure 22.3, are listed below:

- Average compressive strength of M2, (i.e., 20.30MPa) is observed to be 13.9% less when compared with M1 (i.e., 23.60MPa) due to the insufficient angularity of the used artificial aggregate.
- Average compressive strength of M3, (i.e., 22.80MPa) is observed to be 3.3% less when compared with M1 (i.e., 23.60MPa) due to the insufficient angularity of the used artificial aggregate as 50% of coarse aggregate content.
- Average compressive strength of M4 (i.e., 23.4MPa) is approximately similar with compressive strength of M1 (i.e., 23.60MPa); this implies that the replacement of fine aggregate with quarry dust gives same result.
- Average compressive strength of M5 (i.e., 23.1MPa) is approximately equal to the compressive strength of M1 (i.e., 23.60MPa); this implies the bonding between artificial aggregate with quarry dust behaves similar as a bonding between the coarse aggregate with fine aggregate in a concrete.
- Average compressive strength of M6 (i.e., 23.8MPa) is compared with the compressive strength of M1 (i.e., 23.60MPa); this implies that the 50% of fine aggregate replaced with the quarry dust in natural concrete gives the approximate result as natural concrete.
- Average compressive strength of M7 (i.e., 22.10MPa) is observed 6.35% less when compared with compressive strength of M1 (i.e., 23.60MPa), due to using 50% quarry dust and 100% artificial aggregate in a conventional concrete.

The following Conclusions arise from the present study and are listed below:-

- seven trial concrete mixes with artificial aggregates and quarry dust as normal aggregates with cement were prepared and all show nearly same compressive strength as that of conventional concrete, except M2.
- Among all mixes, M2 mix is not showing same strength as conventional because of poor bonding between normal fine aggregates with artificial aggregates.
- An attempt made to replace natural aggregates with quarry dust and artificial aggregates was fulfilled from the results from the Table 22.5 and Figure 22.3.

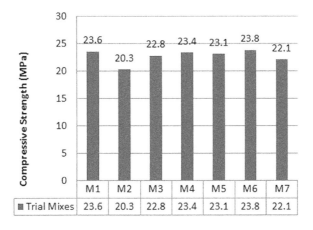

FIGURE 22.3 Comparison of average compressive strength (MPa) of trial mixes at 7 days.

REFERENCES

Baskaran P, Karthick Kumar M, Krishnamoorthy N, Saravanan P, Hemath Naveen K.S. & Vinothan KG, "Partially replacement of fine aggregate with GGBS", *International Journal of Civil Engineering (SSRG-IJCE)*, Vol. 4, (2017), pp. 49–56.

Francesco C. & Raffaele C., "Use of cement kiln dust, blast furnace slag and marble sludge in the manufacture of sustainable artificial aggregates by means of cold bonding pelletization", *Materials Journal*, (2013), 6, pp. 34139–43159.

Harilal B. & Job T., "Concrete made using cold bonded artificial aggregate", *American Journal of Engineering Research (AJER)*, Vol. 1, (2012), pp. 20–25.

Hiraskar K.G. & Chetan P., "Use of blast furnace slag aggregate in concrete", *International Journal of Scientific & Engineering Research*, Vol. 4, (2013), pp. 95–98.

Jayalakshmi S.N. & Basil J., "Study of properties of concrete using GGBS and recycled Concrete cggregates", *International Journal of Engineering Research & Technology (IJERT)*, Vol. 5, (2016), pp. 160–166.

Kavitha, K., Cheela VRS & Raju, SG, "Utilization of quartzite as fine aggregate in concrete". *Journal of Science & Technology*, (2015), 10, No. 5, pp. 45–53.

Narendra M.D. & Gangha G., "An experimental study on high performance concrete partially replacing cement and fine aggregate with GGBS & ROBO sand", *International Journal of Engineering Sciences & Emerging Technologies*, Vol. 7, (2015), pp. 737–742.

Palankar N., Shankar A.R. & Mithun B.M., "Studies on eco-friendly concrete incorporating industrial waste as aggregates", *International Journal of Sustainable Built Environment*, Vol. 4, No. 2, (2015), pp. 378–390.

Pavan Kumar M. & Mahesh Y., "The behaviour of concrete by partial replacement of fine aggregate with copper slag and cement with GGBS", *IOSR Journal of Mechanical and Civil Engineering*, Vol. 12, (2012), pp. 51–56.

Sudheer P., Munireddy M.G., & Adiseshu S., "Study on strength of innovative mortar synthesis with epoxy resin, fly ash and quarry dust", *American Journal of Engineering Research (AJER)*, Vol. 05, No. 05, (2016a), pp. 221–226.

Sudheer P., Munireddy M.G., & Adiseshu S., "An experimental study on strength of hybrid mortar synthesis with epoxy resin, fly ash and quarry dust under mild condition", *Techno Press Journal of Advances in Materials Research, An International Journal (AMR)*, Vol. 05, No. 03, (2016b), pp. 171–179.

Sudheer P., Munireddy M.G., & Adiseshu S., "Study on strength of hybrid mortar synthesis with epoxy resin, fly ash and quarry dust under extreme conditions", *IOP Conference Series: Materials Science and Engineering*, Vol. 225, No. 01, (2017), pp. 012169.

Sudheer P., Munireddy M.G., & Adiseshu S., "Durability study on hybrid mortar under different environmental conditions", *Indian Concrete Institute Journal*, Vol. 19, No. 02, (2018), pp. 35–39.

Ravisankar K., Gowtham S.K. & Raghavan T.R., "Experimental study on artificial fly ash aggregate concrete", *International Journal of Innovative Research in Science and Engineering*, Vol. 4, (2015), pp. 11124–11132.

Sudheer Ponnada, V.R. Sankar Cheela & Gopala Raju S.S.S.V., "Investigation on mechanical properties of composite concrete containing untreated sea sand and quarry dust for 100% replacement of fine aggregate", *Materials Today: Proceedings*, Vol. 32, No. 4, (2020), pp. 989–996. doi:10.1016/j.matpr.2020.06.203.

Shareef U., Cheela V.R.S., & Raju, S.G., "Study on physical and mechanical properties of quartzite and silico-manganese slag as an alternative material for coarse aggregate". *International Journal for Scientific Research & Development*, Vol. 3, No. 9, (2015), pp. 72–74.

Shareef U., Raju S.G., & Cheela V.R.S., "Study on the utilization of quartzite as a replacement for coarse aggregate in concrete", *International Journal of Environment and Waste Management*, Vol. 24, No. 1, (2019), pp. 107–115.

23

Clean C&D Waste Material Cycles through BIM-Enhanced Building Stock Examination Practices: An Austrian Case Study

M. Rašković and A.M. Ragoßnig
RM Umweltkonsulenten ZT GmbH, Austria

K. Kondracki and M. Ragoßnig-Angst
Vermessung Angst ZT GmbH, Austria

CONTENTS

23.1 Introduction 191
    23.1.1 Pre-Demolition Waste Audit 192
    23.1.2 Optimisation Strategy 192
23.2 Objectives 192
23.3 Methodology 192
    23.3.1 Data Capturing 193
        23.3.1.1 Digital Scanning of Spatial Geometry 193
        23.3.1.2 Determination of Material Compositions 193
    23.3.2 Data Modelling 194
        23.3.2.1 Database Assignment and Query 194
23.4 Findings and Result 194
    23.4.1 Innovative 3D Scan-Technologies 194
    23.4.2 Manual as-Built BIM Modelling 195
    23.4.3 Transferability and Efficiency 195
        23.4.3.1 Level of Development 195
        23.4.3.2 Structure Intricacy Dependent Performance 196
23.5 Summary and Outlook 196
References 197

23.1 Introduction

Resource scarcity, sustainability challenges within the construction sector, as well as stricter legislations on the efficient use of resources and on environmentally sound waste management practices instigate companies and organizations in the fields of architecture/engineering/construction (AEC), facility management (FM) and deconstruction to manage resources effectively over the entire lifespan of a building, including its end phase—the demolition stage (Volk et al., 2014). In industrialized countries, the quantity of built-in resources already exceeds the quantity of useful resources occurring in natural deposits (Kovacic et al., 2019). Buildings and infrastructure, therefore, represent valuable material stocks for recovery-oriented dismantling. At the same time, however, in order to guarantee clean material cycles, contained pollutants need to accordingly be discharged and directed into dedicated sinks (European

Parliament & Council, 2004; European Commission, 2015, 2016). For the protection of human health and the environment it is crucial to free accruing demolition waste from any contaminated building components before it undergoes further reuse or recycling procedures.

23.1.1 Pre-Demolition Waste Audit

In Austria, the state-of-the-art approach for identification and controlled separation of potentially contaminated construction waste from valuable building materials, for demolition projects with more than 750 tonnes of demolition waste, is the so-called pre-demolition waste audit; a procedure involving an investigation of pollutants and impurities in buildings prior to their demolition (BMLFUW, 2015; Austrian Standards 2014a, 2014b). One of the many objectives of this conduct, besides the qualitative assessment of waste, is the quantification of waste qualities to be expected in the course of demolition work. Here, building material-related mass estimations of anticipated waste qualities are usually derived from the evaluation of not only on-site but also considerable off-site investigation results. The reliability of the assessment, therefore, significantly depends on the availability of appreciable documentation records. For many existing buildings, however, sufficiently accurate information about their true geometrical structure or their actual material constitution is missing. Situations of this kind lead to mass estimates fraught with high uncertainties, leading to potential cost increases for waste management.

23.1.2 Optimisation Strategy

In order to optimise the steering of material flows, the Austrian engineering consultancy RM Umweltkonsulenten ZT GmbH (hereinafter RMU), as an affiliated partner of the joint research project "SCI_BIM – Scanning and data capturing for Integrated Resources and Energy Assessment using Building Information Modelling" funded by the Austrian Research Promotion Agency (FFG) [Stadt der Zukunft programme, grant number 867314], has developed a potent tool, intended to be provided as a service in the near future. Defined as a mechanism for the quantitative and qualitative documentation of waste qualities and, in particular, localised pollutants in structures, it is able to combine the use of innovative low-tech geometry acquisition systems with a tailor-made transformation process that converts 3D scans into models designed with BIM technologies. In parallel, the company's proprietary web-based geoinformation system (WebGIS) has been refined correspondingly, now supporting centralised model embeddings, digital depositions of chemical quality assessments, as well as automated generations of volume and mass estimates.

23.2 Objectives

Present conference contribution aims to give a concise description concerning the development of a future-oriented software tool for the capturing, the modelling, and the classification of integrated resources in buildings including pollutants and impurities, as well as the presentation of intermediate and final results in the form of acquired research findings and one illustrative depiction of the WebGIS-based end product as a service for pending demolition and disposal planning. In this context, the methodology's transferability from a real case study, used for the very development of the tool, to other types of building structures, which differ in terms of their geometric and structural complexity, is examined in particular.

23.3 Methodology

The process of generating functional, 3D models of existing buildings differs significantly from that for new buildings (Volk, R., et al, 2014). Grossly simplified, the integral process components for creating such as-built BIMs generally include (i) the collection of data providing information about the geometrical and physical constitution (data capturing) and (ii) the creation of digital models based on prepared geometric information (data modelling), which—(iii) through an integration of distinct, project-specific data—should be enabled to provide structured information about the shape and material composition over the entirety of building components (database assignment and query).

Starting point for the three-phase development of a digital prototype for the documentation of preliminary deconstruction planning has been a real case study, embodied by an existing building, which has been used as reference to implement the sub-processes described in points (i) to (iii).

The test object to be surveyed has been a single-story building belonging to the Vienna University of Technology (TU Wien), called "Funkehalle", that is located in the third district of Vienna (see Figure 23.1). Consisting of several office-, seminar- and conference rooms, as well as two machine halls, one server-, one heating- and some sanitary rooms, it covers an area of approximately 1,200 m^2, with an average room height of 3m, whereby the largest and smallest room heights amount to 6.5m and 2.2m. The building's flat roof is formed on three different levels. The gross volume of the test object is rounded to 7,350m^3.

23.3.1 Data Capturing

23.3.1.1 Digital Scanning of Spatial Geometry

In the first phase, geometric data of the test building have sufficiently been recorded by means of various sensor systems operating contactlessly within the optical range of the electromagnetic spectrum (Wiedemann, 2004). These records, also known as scans, subsequently—if appropriate—have been evaluated for each system individually, yielding a three-dimensional point cloud (Otepka et al., 2013) in each reasonable case. At this project stage, technological limits were identified, and the recording systems' achievable accuracies were examined by comparison with traditional, geodesic engineering measurement systems providing high resolution. At the same time, the suitability of the applied low-tech technologies and methods for an efficient as-built BIM creation process was tested.

23.3.1.2 Determination of Material Compositions

In the same development phase, an initial pre-demolition waste audit has been carried out by a competent person, in order to determine the material compositions of the test building's components. Generalised, the concrete procedure comprises the following steps: 1. Research with regards to documentations on structure, location and use; 2. Investigation of pollutants and impurities listed in the Austrian Standard ÖNORM B 3151 (Austrian Standards, 2014b) in all areas of the structure, review and verification of the research findings, registration and documentation of new information; 3. Sample extraction in areas suspected to be contaminated; 4. Chemical analyses; 5. Assessment of waste qualities and estimation of their volumes and masses expected in the course of the upcoming demolition work; 6. Documentation of the

FIGURE 23.1 Aerial photograph (Vienna GIS, 2019) (left) and of the test object to be surveyed (right).

results—including photos, planning documents and inspection reports based on chemical analyses—and investigation report.

23.3.2 Data Modelling

On the basis of registered point clouds—which, as a whole, are to be interpreted as a discretisation of individual surfaces of registered building components—a geometric model, consisting of interlocking building elements, was created manually using Graphisoft's BIM modelling software ArchiCAD in the second phase of this project.

23.3.2.1 Database Assignment and Query

In the third and final phase, the geometry model generated in the previous step was used to further develop the company's own WebGIS in terms of centralised model embeddings, digital deposits of chemical quality assessments, and automated generations of volume and mass estimates of expected waste qualities. The results of the latter are, *inter alia*, collectedly expressed in the form of downloadable Excel sheets, which, in addition, contain treatment classifications subordinated to the identified waste qualities. Therein, waste categories and qualities are expressed in terms of identification numbers specified by the government (BMLFUW 2003; Austrian Standards 2005), which, for the automatic generation of volume and mass estimates, were drawn from an already prepared but adaptable register of relevant waste codes.

The geometric BIM model was first loaded into a cloud using a web-based application programming interface (API), both provided by Autodesk. Simultaneously, the RMU WebGIS was functionally extended to embed the model visualisation in the browser window, where the uploaded and embedded BIM model could interactively be enriched with information about the layered building materials resulting from the previously performed pre-demolition waste audit using other tools from the same API platform (Autodesk Forge). Through an additional implementation of suitable calculations processes, which—on the one hand—are systematically referenced to a defined but expandable list with tailored parameters, like material density and surface weight constants, and which—on the other hand—filter dimensional information of individual building components already contained in the geometry model, an automated calculation of the volumes and masses of various waste qualities was made possible. Process workflows for uploading and building component layer-related linking of relevant photos, digital planning documents and chemical analyses were also integrated.

23.4 Findings and Result

23.4.1 Innovative 3D Scan-Technologies

For the specific use case, it became apparent that—when it comes to handling, time efficiency and overall performance—hand-held 3D laser scanners available on the market today, which work according to the principle of Simultaneous Localisation And Mapping (SLAM) (Nüchter, 2009) in ranges of up to 30m, are largely superior to so-called compact laser scanners, which are stationary but dynamic sensor systems, thanks to their low weight. Nevertheless, terrestrial compact laser scanners were additionally used as supplements in exterior areas (i.e., for capturing facades or roofs) where former systems generally have difficulties with regards to the automated merging of single scans. In view of the geometric complexity of the building's interior, the initial idea of surveying the test object photogrammetrically had to be rejected. A drastic increase of captured images for a sufficient coverage of the building's geometry would have led to an immense increase in both, acquisition, and evaluation time. Unfortunately, it was also necessary to desist from innovative low-tech ranging technologies potentially useful for sufficient building surveying, such as depth cameras or smart glasses working according to the principle of active stereoscopy. In-house tests have proven their inefficiency and have shown to be still in their infancy when it comes to large-scale surveying projects.

23.4.2 Manual as-Built BIM Modelling

This type of reconstruction process (Scan-to-BIM, Points-to-BIM), which is geared to project-specific requirements with regards to the Level of Geometry (LoG), has proven its worth in practice so far—despite the great amount of time required. The main reason for favouring manual processing lies in the fact that automated or semi-automated methods, which extensively have been dealt with in research and development, are not (yet) capable of comprehensively considering constructional characteristics or semantic information (Volk et al., 2014). Nonetheless, within the framework of the project SCI_BIM, measures are being developed to circumvent these shortcomings, paving the way for automated and coherent BIM model creation processes.

More importantly, when it comes to manually reconstructing BIM models from point clouds, it is crucial to consider the surveying system's underlying operating principles. 3D laser scanners, as well as photogrammetric or depth cameras, are non-invasive imaging instruments detecting ranges to object elements without penetrating them. Hence, if a building's foundations are to be modelled adequately, additional information is necessary, which usually is drawn from old construction plans.

23.4.3 Transferability and Efficiency

The transferability of the methodology from the real case study to buildings of deviating constructional complexity is essentially feasible and is fundamentally subject to the same conditions that apply to the test object. However, it is generally accompanied by constraints (or even benefits), having corresponding impacts on the overall efficiency of the tool.

In principle, the overall efficiency of the tool is the sum of its dynamic sub-efficiencies, which depend on the amount of information taken from different categories. In general terms, these partial efficiencies refer to the following four consecutive subtasks: the collection of, the modelling of, the model assignment with, and the model query about information of geometric and/or material nature. In contrast to the last three operations, the preliminary building data acquisition is two-tiered.

However, for the comparability in the transfer of the methodology from the real case study to buildings of deviating structural complexity, the one-time effort for extending the RMU WebGIS is not to be considered, since it represents the same constant factor in the course of determining the overall efficiencies. The same applies to the workload for querying the building database, which is negligibly small compared to the expenditure of the remaining project sub-tasks and thus does not significantly affect the determination of the overall efficiency.

23.4.3.1 Level of Development

The partially interdependent, individual efficiencies of the two-tiered building data acquisition, the building data modelling and the building database assignment inevitably depend on the initial, project-specific determination of the Level of Development (LoD) of the model and its components. As an indicator for the model's reliability, it is affected by both, the degree of geometric detail (LoG) and the information depth (Level of Information – LoI). It generally applies that LoD = LoG + LoI (BIM Forum, 2019).

Optimal project assumptions for the LoG as well as for the LoI significantly depend on the extent of optically visible surfaces of building components to be captured, as well as on the complexity of the building's spaciousness (spatial irregularities such as anglings or segmentations) and on the building's structural realisations (such as basic construction types and layer structures).

It ought to be noted at this point that both, LoG and LoI are values bounded from above due to the diminishing marginal benefit occurring induced by a rising model granularity. This implies that LoD, as well, must be bounded from above and cannot be increased endlessly by simply boosting one or both of its (actually limited) summands as far as possible. That is to say, the formula given above could also be rewritten to LoD = LoG*LoI, whereby LoG and LoI are representing saturation levels ranging from 0 to 1.

FIGURE 23.2 RMU WebGIS browser view of the fully implemented test building model.

23.4.3.2 Structure Intricacy Dependent Performance

The smaller or geometrically more complex the structural components to be scanned, the lower the recording distance and the higher the sampling interval of the laser, or rather, the higher the number of recording positions must be selected. Demands on the geometric resolution of this type lead to an extension of the trajectory of recording positions, which, in turn, lead to an extension of the recording time and thus to an increase in the accumulated geometric data set. This eventually increases the cost of capturing and subsequent processing of recorded information to a geometric as-built BIM model. The same applies to the requirement imposed on the material information density. The more unrestricted the actual bandwidth and distribution of the materials installed, the higher the outlay for its on-site exploration and the resulting effort for its utilisation—the model enrichment.

Thus, for both, the LoG and LoI a directly proportional relation to its respective sub-effort, and at the same time, an indirectly proportional relation to its partial efficiency is given. However, since both summands of the LoD are completely independent of one another, the overall effort, or rather, the overall efficiency of the tool cannot be given in the form of a benchmark (such as the time required per building volume) which could be then used for linear extrapolation in order to determine the methodology's efficiency when transferring the methodology to different types of building complexes.

For the sake of completeness, the times required in each of the four significant project sub-tasks until the completion of the tool on the basis of the test object (see Figure 23.2) should be given nonetheless: Recording and processing of geometric information—5 hours, capturing and processing of building material information—19 hours, geometric modelling—40 hours, database set up—5 hours.

23.5 Summary and Outlook

In order to optimise the steering of stock material flows, the Austrian engineering consultancy RM Umweltkonsulenten ZT GmbH created a powerful, serviceable tool, which, in the foreseeable future, will be integrated as a technological component into existing pre-demolition waste audits, responsible for qualitatively more transparent and quantitatively more precise documentations of waste qualities. In the course of its development, the company's proprietary web-based Geographic Information System was appropriately complemented by functionalities and process flows in order to deal with extensive visual and numerical condition assessments of demolition buildings. With the related achievable reliability improvement of demolition waste mass estimations, prospective beneficial impacts, such as accurate calculations and minimisations of disposal costs, landfill volume conservations, exploitations of secondary resources to an extremely high degree and an ongoing maintenance of clean construction and demolition waste material cycles, shall be ensured, contributing to the good of humanity and the environment.

REFERENCES

Austrian Standards (2005). Abfallverzeichnis (List of wastes), ÖNORM S 2100: 2005-10-01.

Austrian Standards (2014a). Innenraumluftverunreinigungen – Teil 32: Untersuchung von Gebäuden auf Schadstoffe (Indoor air – Part 32: Investigation of buildings for the occurrence of pollutants), ÖNORM EN ISO 16000-32: 2014-10-01.

Austrian Standards (2014b). Rückbau von Bauwerken als Standardabbruchmethode (Dismantling of buildings as a standard method for demolition), ÖNORM B 3151: 2014-12-01.

BIMForum (2019). LoD specification for building information models part I. Retrieved October 1, 2019, from https://bimforum.org/wp-content/uploads/2019/04/LOD-Spec-2019-Part-I-and-Guide-201 9-04-29.pdf.

BMLFUW (2003). Bundesgesetz BGBl. II Nr. 570/2003 des Bundesministers für Land- und Forstwirtschaft, Umwelt und Wasserwirtschaft über ein Abfallverzeichnis, zuletzt geändert durch das Bundesgesetz BGBl. Nr. 498/2008 (Federal Law BGBl. II No. 570/2003 of the Federal Ministry for Agriculture, Forestry, Environment and Water Management on a list of wastes as last amended by BGBl. II No. 498/2008).

BMLFUW (2015). Bundesgesetz BGBl. II Nr. 181/2015 des Bundesministers für Land- und Forstwirtschaft, Umwelt und Wasserwirtschaft über die Pflichten bei Bau- und Abbruchtätigkeiten, die Trennung und die Behandlung von bei Bau- und Abbruchtätigkeiten anfallenden Abfällen, die Herstellung und das Abfallende von Recycling-Baustoffen, zuletzt geändert durch das Bundesgesetz BGBl. II Nr. 290/2016 (Federal Law BGBl. II No. 181/2015 of the Federal Ministry for Agriculture, Forestry, Environment and Water Management on duties in relation to construction and demolition activities, separation and treatment of waste incurred from construction and demolition activities, production and end-of-waste criteria for recyclable building materials as last amended by BGBl. II No. 290/2016).

European Commission (2015). Closing the loop – An EU action plan for the circular economy, COM (2015) 614 final, p. 25. Retrieved October 1, 2019, from https://eur-lex.europa.eu/resource.html?uri=cellar:8a8ef5e8-99a0-11e5-b3b7-01aa75ed71a1.0004.02/DOC_1&format=PDF.

European Commission (2016). Commission Regulation (EU) 2016/640 of 30 March 2016 amending Annexes IV and V to Regulation (EC) No 850/2004 of the European Parliament and of the Council on Persistent Organic Pollutants. *Official Journal L* 80, pp. 17–24. Retrieved October 1, 2019, from https://eur-lex.europa.eu/legal-content/DE/TXT/PDF/?uri=CELEX:32016R0460&from= DE.

European Parliament & Council (2004). Regulation (EC) No 850/2004 of the European Parliament and of the Council of 29 April 2004 on Persistent Organic Pollutants and Amending Directive 79/117/EEC. *Official Journal L* 158, pp. 7–49. Retrieved October 1, 2019, from https://eur-lex.europa.eu/legal-content/DE/TXT/PDF/?uri=CELEX:32004R0850&from=en.

Kovacic, I., Honic, M., Rechberger, H., Oberwinter, L., Lengauer, K., Hagenauer, A., Glöggler, J., & Meier, K. (2019). Prozess-Design für den "Building Information Modeling" (BIM) basierten, materiellen Gebäudepass – BIMaterial (Process-design for a BIM-based material pass-port – BIMaterial). Berichte aus Energie- und Umweltforschung Schriftenreihe 8/2019, herausgegeben vom Bundesministerium für Verkehr, innovation und technologie (BMVIT). Retrieved October 1, 2019, from https://nachhaltigwirtschaften.at/resources/sdz_pdf/schriftenreihe-2019-8-bimaterial.pdf.

Nüchter, A. (2009). *3D Robotic Mapping*. Berlin, Germany: Springer-Verlag

Otepka, J., Sajid, G., Waldhauser, C., Hochreiter, R., & Pfeifer, N. (2013). Georeferenced Point Clouds: A Survey of Features and Point Cloud Management. *ISPRS International Journal of Geo-Information* vol. 2, no. 4, pp. 1038–1065.

Vienna GIS (2019). Geodatenviewer der Wiener Stadtvermessung MA 41. Retrieved October 1, 2019, from https:// www.wien.gv.at/ma41datenviewer/public/.

Volk, R., Stengel, J., & Schultmann F. (2014). Building Information Models (BIM) for Existing Buildings – Literature Review and Future Deeds. *Automation in Construction* 38, pp. 109–127.

Wiedemann, A. (2004). *Handbuch Bauwerksvermessung – Geodäsie, Photogrammetrie, Laserscanning (Handbook Building Survey – Geodesy, Photogrammetry, Laserscanning)*. Basel, Switzerland: Springer AG.

24

Fly Ash-Based Jute Fiber Reinforced Concrete: A Go Green Approach for the Concrete Industry

Bidhan Ghosh and T. Senthil Vadivel
Adamas University, Kolkata, India

CONTENTS

24.1 Introduction 199
24.2 Background of the Study 199
24.3 Problem Description 200
24.4 Experimental Program 200
24.5 Conclusion 205
References 205

24.1 Introduction

Concrete is the most extensively used construction material in the world and India is the second largest user of it. Concrete technologists are continuously carrying out research to improve the rheological and mechanical properties deliberated to be used in a particular structure, its environment and construction methodology to meet the functional, strength, economy, and durability requirements. The American Concrete Institute defines HPC as concrete which meets special performance requirements and behavioral uniformity that cannot always be achieved routinely by using only conventional materials and normal mixing, placing and curing practices (ACI 363R 1992). Meeting the target performance specification and performance uniformity of HPC is closely linked with concrete production, transportability, rheological and hardened properties such as placement, compaction without segregation, long term mechanical properties, early age strength, toughness, permeability, density, heat of hydration, volume stability, or service life in severe environments. The disadvantage of normal cement concrete is its brittleness, with relatively low tensile strength and poor resistance to crack generation and propagation and a very little elongation at breaking point. One among the author already tried with waste plastic as fiber in concrete and found improvement in compressive strength and now an attempt made to use natural fibers in concrete to prevail over these discrepancies in concrete. (Senthil Vadivel & Doddurani, 2013). The specific solution lies in incorporation of short, discrete, dispersed natural bast fiber which might play an important role to improve rheological and mechanical properties of concrete.

24.2 Background of the Study

A wide range of literature has been reviewed. A list of a few references is appended at the end of the chapter. The literature survey covers primarily the following areas:

- Plastic and hardened characteristics of concrete.
- Concrete characteristic evaluation.

DOI: 10.1201/9781003217619-27

- Performance of concrete with jute fiber.
- Impact of ingredients on concrete properties.
- High Performance Concrete (HPC).
- Microstructure of concrete.

Mansur (1981) investigated concrete and mortar having jute fibers with various lengths as reinforcements, which were randomly oriented and uniformly presented in the matrix. The outcome of the experiment revealed that jute fibers are appropriate for a low-cost construction material. Razmi & Mirsayar (2017) studied that there is great improvement in compressive, tensile, and flexure strength of jute fiber concrete. The mode of fracture toughness is also enhanced by using jute fibers in the concrete mixture. Ramakrishna & Sundararajan (2005) reported that different natural fibers like coir, sisal, jute, hibiscus cannebinus, in concrete shows improvement in ductility, impact and fracture toughness, and also reduction of shrinkage. Zakaria et al. (2016) confirmed from the test result that the length of jute fiber has an effect on the mechanical properties of concrete. Warke & Dewangan (2016) compared the performance, workability, and strength of jute fiber concrete with plain concrete and demonstrated the scope of using jute fiber to improve concrete strengths and performance.

24.3 Problem Description

Natural (cellulosic) fibers might offer the opportunity as a convenient reinforcing agent in concrete composite due to its low density and high tensile property. Two natural (cellulosic) bast fibers, namely Olitorius and Capsularies, are taken as jute reinforcement. Three different types of lengths of each fiber are chosen as 10mm, 15mm and 20mm. A total of 8 values of jute percentages were incorporated in cement concrete as 0.25% to 2%, with an increment of 0.25% for each jute fiber. Various specimens were made considering different lengths and different percentage of jute for each fiber. Cubes specimens were tested to determine the compressive strength of concrete on 7, 14 and 28 days. The different major properties of all the materials including jute fibers were also tested, and the results were verified.

24.4 Experimental Program

Two types of jutes, namely Olitorius and Capsularies, were taken and tested from the Central Research Institute for Jute and Allied Fibers, Kolkata. The fineness of fiber and tensile strength are shown in Tables 24.1 and 24.2.

Portland pozzolana cement (fly ash-based) conforming to IS: 1489 (Part-I)-1991 was used. Its physical and chemical properties are shown in Tables 24.3 and 24.4.

Naturally available river-sand is used in this project as fine aggregate. All the tests are conducted as per Indian Standard Specifications IS: 383. The physical properties, i.e., sieve analysis, water absorption, and specific gravity are shown in the Tables 24.5 and 24.6.

Adequately cleaned dry stones are used as a coarse aggregate for all the mix in this project. All the tests were conducted as per IS 383 Indian standard code. The physical properties, i.e., sieve analysis, water absorption, and specific gravity are shown in the Tables 24.7 and 24.8.

The mix design was done as per IS 10262. The target strength was M40 with standard deviation of 5. Adopted w-c ratio is taken as 0.45 with target slump of 100mm. Per cubic meter of concrete, the weight of cement, fine aggregate, coarse aggregate and water required were 438kg, 659.8kg, 1,136kg and 197liters, respectively. Two natural bast fibers, namely Olitorius and Capsularies, were taken as jute reinforcement. Three different lengths of each fiber were chosen as 10mm, 15mm and 20mm. A total of eight values of jute percentage were incorporated in cement concrete as 0.25%, 0.5%, 0.75%, 1%, 1.25%, 1.5%, 1.75% and 2%. The compressive strength test was conducted to determine the characteristic compressive strength of concrete with surface dry condition after 7, 14 and 28 days of curing as per IS: 516-1959.

TABLE 24.1

Fineness of Fibers

	Fineness of Fiber	
S. No.	Fiber weight (g)	Fineness
	Fiber: Olitorius	
1	3.0039	2.4
2	3.0017	2.4
3	3.0002	2.4
4	3.0043	2.4
5	3.0028	2.2
	Fiber: Capsularis	
1	3.0028	0.9
2	3.0029	0.9
3	3.0038	1
4	3.0032	0.9
5	3.0017	0.9

TABLE 24.2

Tensile Strength of Fibers

	Fiber Strength		
S. No.	Fiber length	Fiber weight (mg)	Tensile Strength (N/mm2)
		Fiber: Olitorius	
1	12.5	298	18.2
2	12.5	294	14.6
3	12.5	301	13.4
4	12.5	295	21.6
5	12.5	303	10.8
		Fiber: Capsularis	
1	12.5	305	10.9
2	12.5	314	5.1
3	12.5	320	10.4
4	12.5	308	4.8
5	12.5	310	7.9

TABLE 24.3

Physical Properties of Portland Pozzolana Cement (PPC)

S. No	Characteristics	Experimental Value	As per IS:1489 (Part-I) 1991
1	Normal Consistency	33 min	34 min
2	Specific Gravity	3.13	3.15
3	Initial Setting Time	140 min	30 min, Min
4	Final Setting Time	525 min	600 min, Max
5	Soundness of Cement	6mm	10mm
7	Compressive Strength 7 days	28 Mpa	22 Mpa, Min
	28 days	44 Mpa	33 Mpa, Min

TABLE 24.4

Chemical Properties of Portland Pozzolana Cement

S. No.	Characteristics	Value	As per IS:1489 (Part-I) 1991
1	% Magnesia	1.1	6 max
2	% Sulphur	2.3	3 max
3	% Loss on Ignition	2	5 max
4	% Chloride	0.022	0.1

TABLE 24.5

Sieve Analysis of Fine Aggregate

IS Sieve (mm)	Weight Retained on Sieve (gm)	% Retained	Cumulative % Retained	% Passing	As per IS: 383 (zone IV)
4.75	0	0	0	100	95–100
2.36	1	0.2	0.2	99.8	95–100
1.18	5	1	1.2	98.8	95–100
0.600	26	5.2	6.4	93.6	90–100
0.300	186	37.2	43.6	56.4	80–100
0.150	267	53.4	97	3	15–50
Pan	15	3	100	0	0–15
			Σ=248.4		

Fineness Modulus = 2.48 (Zone IV)

TABLE 24.6

Physical Properties of Fine Aggregate

S. No.	Characteristics	Experimental Value
1	Specific Gravity	2.5
2	Fineness Modulus	2.48 (Zone IV)
3	Water Absorption	2%

TABLE 24.7

Sieve Analysis of Coarse Aggregate

IS Sieve (mm) (mm)	Weight Retained on Sieve (gm)	% Retained	Cumulative % Retained	% Passing	As per IS: 383 (20 mm)
80	0	0	0	100	–
63	0	0	0	100	–
40	0	0	0	100	100
20	127	6.35	6.35	93.65	85–100
16	515	25.75	32.1	67.9	–
12.5	662.5	33.125	65.225	34.775	–
10	448.5	22.425	87.65	12.35	0–20
4.75	214	10.7	98.35	1.65	0–5
2.36	33	1.65	100	0	–
1.18	0	0	100	0	–
0.600	0	0	100	0	–
0.300	0	0	100	0	–
0.150	0	0	100	0	–
			Σ=789.675		

Fineness Modulus = 7.89 (maximum aggregate size = 20mm)

TABLE 24.8
Physical Properties of Coarse Aggregate

S. No.	Characteristics	Experimental Value
1	Specific Gravity	2.75
2	Fineness Modulus	7.89 (maximum aggregate size = 20mm)
3	Water Absorption	0.5%

Cube size of 150mm × 150mm × 150mm cubes were used for determining the compressive strength of concrete. The results of compressive strength of fiber reinforced concrete (FRC) at 7 days, 14 days and 28 days are plotted in Figure 24.1 to Figure 24.6.

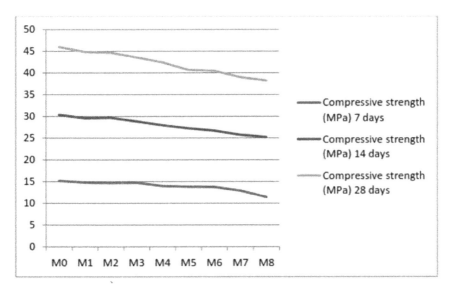

FIGURE 24.1 Compressive strength of Olitorius jute FRC of 10mm length.

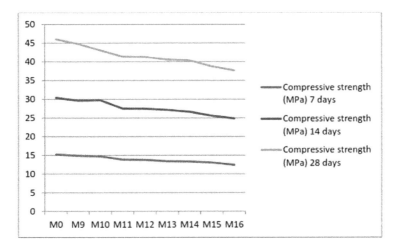

FIGURE 24.2 Compressive strength of Olitorius jute FRC of 15mm length.

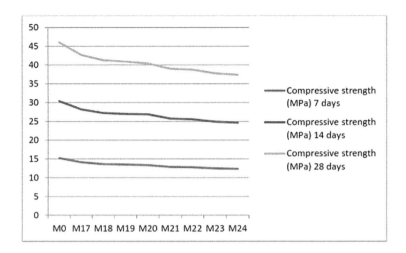

FIGURE 24.3 Compressive strength of Olitorius jute FRC of 20mm length.

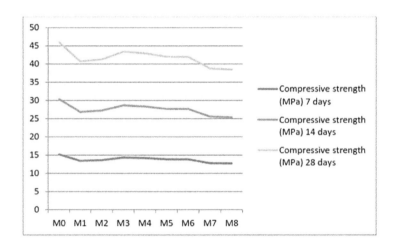

FIGURE 24.4 Compressive strength of Capsularies jute FRC of 10mm length.

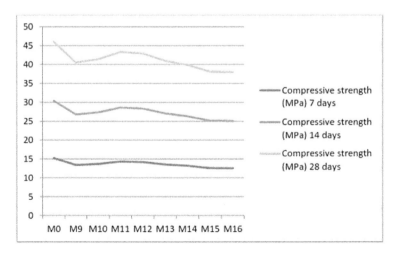

FIGURE 24.5 Compressive strength of Capsularies jute FRC of 15mm length.

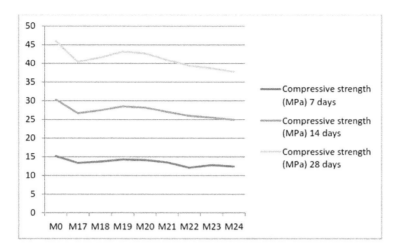

FIGURE 24.6 Compressive strength of Capsularies jute FRC of 20mm length.

24.5 Conclusion

- It was observed, while using Olitorius fiber in concrete, that the compressive strength decreases when the percentage of jute increases. The maximum compressive strength achieved in this fiber is at 0.25% of jute.
- The good increment in compressive strength was found while using Capsularies fiber in concrete and the maximum compressive strength was achieved in 0.75% of jute.
- In general, compressive strength was decreased while increasing the length of the jute fiber.

REFERENCES

Mansur, M.A. (1981) "Jute Fibre Reinforced Composite Building Materials," Proc. 2nd Australian Conf. Eng. Materials, University of New South Wales, 1981, pp. 585–596.

Ramakrishna, G., & Sundararajan, T. (2005). "Impact strength of a few natural fibre reinforced cement mortar slabs: A comparative study", *Cement and Concrete Composites*, 27, 547–553. doi: 10.1016/j.cemconcomp.2004.09.006.

Razmi, Ali & Mirsayar, Mirmilad. (2017). "On the mixed mode I/II fracture properties of jute fiber-reinforced concrete", *Construction and Building Materials*, 148, 512–520. doi: 10.1016/j.conbuildmat.2017.05.034.

Senthil Vadivel, T. and Doddurani, M. (2013), An experimental study on mechanical properties of waste plastic fiber reinforced concrete, *International Journal of Emerging Trends in Engineering and Development*, Vol. 2 (3). Available in online: http://www.rspublication.com/ijeted/ijeted_index.htm

Warke, P., & Dewangan, S. (2016), "Evaluating the Performance of Jute Fiber in Concrete", *International Journal of Trend in Research and Development*, 3(3), ISSN: 2394-9333.

Zakaria, M., Ahmed, M., Hoque, M.M. and Islam, S. (2016). "Scope of using jute fiber for the reinforcement of concrete material". *Textiles and Clothing Sustainability*, 2. doi: 10.1186/s40689-016-0022-5.

Part IV

Energy Recovery and Resource Circulation in Construction

25 A Study on Tensile Strength and Modulus Properties of Concrete Using Industrial Waste Vermiculite and Granite-Fines

Damodhara Reddy Budda, Kiran Kumar Narasimhan, and Jyothy S. Aruna
S.V. University College of Engineering, Tirupati, India

C. Sasidhar
JNTU, Andhra Pradesh, India

CONTENTS

25.1 Introduction 209
    25.1.1 General 209
    25.1.2 Present Work 210
    25.1.3 Research Objectives 210
    25.1.4 Review of Literature 210
25.2 Materials and Methods 211
    25.2.1 Materials 211
    25.2.2 Methods 212
        25.2.2.1 Split Tensile Strength 212
        25.2.2.2 Test on Modulus of Elasticity 212
25.3 Research Methodology 212
25.4 Results and Discussion 213
    25.4.1 General 213
    25.4.2 Effect of Granite Fines on Split Tensile Strength 213
    25.4.3 Modulus of Elasticity 213
    25.4.4 X-Ray Powder Diffraction Analysis 214
25.5 Conclusion 216
References 217

25.1 Introduction

25.1.1 General

There is no doubt that with the development of human civilization, concrete will continue to be the dominant construction material in the future. Unfortunately, depleting levels of natural sand is forcing us to search for alternatives for natural sand. Recently, Andhra Pradesh government has imposed restrictions on sand removal from river-beds. As the cost of sand is increasing, the need for focus on an alternative cheaper material is gaining its prominence. Around 18 million tonnes of granite powder is produced by Indian Granite Industry and these are accumulating year by year and causing environmental pollution. In the present investigation, an attempt is made in the direction of choosing an alternative to natural sand in the form of crushed granite powder.

25.1.2 Present Work

In the present study, we cast cement concrete cubes and cylinders with 0%, 10%, 20%, 30% and 40% replacement of sand with granite fines. The method adopted in the investigation was used as per the code specifications of IS 456: 2000, and IS 2386. The water/cement ratio adopted is 0.5, with mix design ratios of water: cement: sand: gravel of 0.5:1:1.73:3.3, respectively. The mix design was carried out using IS 10262: 2009.Tests were carried out at 28 days of water curing.

25.1.3 Research Objectives

1. To determine the degree of tensile strength improvement in concrete obtained after addition of granite powder at proportions 0%, 10%, 20%, 30% and 40%.
2. To determine the modulus of elasticity in concrete obtained after addition of granite powder at proportions 0%, 10%, 20%, 30% and 40%.
3. To determine properties of granite-fines concrete using XRD testing.

25.1.4 Review of Literature

Olawvyi et al. (2012), during their study on the effect of replacing sand with granite fines on the compressive and tensile strength of palm kernel shell concrete, also revealed that 100% of replacement of natural sand by crushed granite fines is possible, and tensile strength increases with increase in crushed granite fines and with the curing age.

Bahoria et al. (2013) opined that natural sand can be replaced partially with waste products like rock flour (40%) and also recommended usage of 20% replacement of natural sand by crushed granite fines when laying rigid concrete pavements.

Prince Arulraj et al. (2013) opined that for M30 and M40 grade concretes with 15% replacement of sand by granite powder resulted in optimum results. For grades M15, M20, and M25, the replacement levels are slightly greater, and M30 and M40 grade concretes with 20% replacement of sand by granite powder resulted in optimum results.

Hameed &Sekar (2009) studied the applicability of quarry dust and marble powder as possible replacements for natural sand in concrete with 50% replacement of sand by marble sludge and quarry rock dust in higher split tensile strength. They have also emphasized that with decrease in the content of crusher dust, split tensile and flexural strength increased.

BaharDemitrel (2010) studied the effect of the using waste marble dust as replacement for fine-sand at proportions of 0%, 25%, 50% and 100% by weight at curing ages of 3, 7, 28 and 90 days and the compressive strength and dynamic modulus of elasticity increases at particular properties between 25% and 50% on the mechanical properties of concrete.

Shirulea et al. (2012) studied partial replacement of cement with marble dust powder revealing that the split tensile strength of concrete in which cement is partially replaced with marble dust powder resulted in increased compressive strength at 10%.

Raman et al. (2007),during their study on non-destructive evaluation of flowing concretes incorporating quarry waste, explained that dynamic modulus of elasticity varies linearly with compressive strength.

Uchikawa&HaneharaHirao (1996),when studying theinfluence of microstructure on the physical properties of concrete prepared by substituting mineral powder for part of aggregate, revealed that with addition of mineral powder the viscosity of concrete increased, and fluidity of concrete reduced. Young's modulus decreased because of fractures in cement paste and dynamic Young's modulus was only half that of aggregate.

Safiuddin et al. (2007), during their study about utilization of quarry waste as fine-aggregate in concrete mixtures, revealed that the dynamic modulus of elasticity and slump increased marginally, and concluded that quarry dust can be used as replacement for natural sand.

A Study on Tensile Strength and Modulus Properties 211

According to Kanmalai William et al. (2008), during their study on mechanical properties of high performance concrete incorporating granite powder as fine-aggregate, has opined that modulus of elasticity for both 7 and 90 days show higher values for all proportions of 25%, 50%, 75% and 100% of replacement of sand with granite fines, compared to 0% replacement of sand. However, he disclosed that a 25% replacement level resulted in a higher modulus of elasticity when compared with other proportions.

25.2 Materials and Methods

25.2.1 Materials

The physical-chemical properties of cement, sand, gravel made of vermiculite and water used in the investigation were analyzed based on, and also using, the standard experimental procedure laid down in the standard code, IS 2386 (Part1): 1963. These standard experimental procedures were adopted for the determination of split tensile strength test and modulus of elasticity of cement concrete. The materials used in the experimental investigation include ordinary Portland cement., granite fines, fine aggregate, coarse aggregate, and water.

The granite fines are used to replace the natural sand in our present sand. The granites are obtained from a stone crushing unit located near KongareddiPalli, Chittoor. The physical and chemical properties of granite fines are listed in Table 25.1 and Table 25.2.

TABLE 25.1

Range of Properties of Granite

S.No	Properties	Values
1	Water Absorption	1 %
2	Specific Gravity	2.6
3	Crushing Strength	2,500kg/m^2
4	Color	Mostly Light Colored
5	Fineness Modulus	2.35

TABLE 25.2

Composition of Granite

Composition Name	Percentage Present
SiO_2	72.04%
Al_2O_3	4.12%
K_2O	4.12%
Na_2O	3.69%
CaO	1.82%
FeO	1.68%
Fe_2O_3	1.22%
MgO	0.71%
TiO_2	0.30%
P_2O_5	0.12%
MnO	0.05%

25.2.2 Methods

25.2.2.1 Split Tensile Strength

As per IS specifications IS 516: 1999, split tensile test is to be conducted. The cylinder of size is 150mm diameter and 300mm length is made and cylinders were placed horizontally between the loading plate surfaces of a compression testing machine. Then the load was applied to the cylinder until failure of the vertical diameter along the cylinder length. The maximum load on the cylinder was to be observed.

$$\sigma_{ct} = 2P/\pi dl \qquad (25.1)$$

Where
σ_{ct} = indirect tensile strength,
d = diameter of cylinder in mm,
l = length of cylinder in mm and
P = maximum load applied to the cylinder

25.2.2.2 Test on Modulus of Elasticity

Fix both top and bottom frame by positioning the spacers and position the pivot rod on screws in locked position and keep tightening screws of bottom and top frame unscrewed. Position the specimen on flat surface, Place the compressometer on the center of the specimen such that tangent screws of both the frames are at equal distance. Later tighten screws and by unscrewing the spacer screws remove spacers.

Specimen assembly is placed in the compression testing machine at center and load is applied at a rate of 140kg/cm²/minute until a stress of (1 + 5)kg/cm² is reached where 1 is one third of the average compressive strength of cubes calculated to the nearest 5kg/cm² (a load of 12.4 tonnes) and for one minute the load is maintained and then gradually reduced to an average stress of 1.5kg/cm² (a load of 0.3 tonnes). The load is applied again at the same rate until an average stress of (c + 1.5) kg/cm² is reached (a load of 11.8 tonnes) and the compressometer reading at this load is taken. Reduce the load gradually to 0.3 tonnes and take the readings. Repeat this process for a third time and take the values.

Young's modulus or tensile modulus can be expressed as

$$E = \text{stress/strain} = (F/A)/(dL/L) \qquad (25.2)$$

where
E = Young's modulus (N/m²) (lb/in², psi)
F = force (N) (lb)
A = area of object (m²) (in²)
dL = elongation or compression (offset) of the object (m) (in)
L = length of the object (m) (in)

25.3 Research Methodology

The following research methodology is adopted (Table 25.3).

1. Collection of materials for preparation of concrete like cement, sand, gravel, granite fines, and water and carrying out preliminary tests on materials and fresh concrete.
2. Using Molds, concrete cubes and cylinders are cast, based on mix design and cured for the specified 28 days.

TABLE 25.3

Test Specimens

Material Properties at 28 days	Shape and Dimensions of Specimens	No of Specimens
XRD test	Cube 150mm × 150mm × 150mm	9
Split Tensile Strength	Cylinder 150mm × 300mm	15
Modulus of Elasticity	Cylinder 150mm × 300mm	15

3. Cured samples are tested for split tensile strength, and modulus of elasticity. Cured samples tested for resisting acid are soaked in acid for 32 days after 28 days water curing, and later tested for XRD test.

25.4 Results and Discussion

25.4.1 General

The results of the present investigations are presented both in tabular and graphical forms. In order to facilitate the analysis, interpretation of the results is carried out at each phase of the experimental work. This interpretation of the results obtained is based on the current knowledge available in the literature as well as on the nature of results obtained. The significance of the result is assessed with reference to the standards specified by the relevant IS Codes. Test results of split tensile strength, and modulus of elasticity are displayed.

25.4.2 Effect of Granite Fines on Split Tensile Strength

The split tensile strength values in N/mm^2 at curing period of 28 days reveal that GF10, GF20, and GF40 have given 1.698, 1.98, 1.28 against a reference mix, GF0, whose value is 2.265. Only GF30 has given a value 2.27, which is slightly greater than the value of reference mix GF0. The values obtained are shown in the Figure 25.1 in which reference line for GF0 and values of GF10,GF20,GF30 and GF40 are shown clearly. It is evident from the graph that only GF30 has crossed the split tensile strength of that of _reference mix GF0 slightly. These values are in accordance with the study carried out by Kanamalai Williams et al.

25.4.3 Modulus of Elasticity

The modulus of elasticity value for 0% is 49,006N/mm^2, for 10% is 38,215N/mm^2, for 20% is 49,249N/mm^2, for 30% is 51,000N/mm^2, and for 40% it is 16,984N/mm^2, respectively. The modulus of elasticity data is plotted. The measurements were performed at age of 28 days. The tests show that modulus of elasticity of all the concrete mixtures were almost a bit higher than reference mix except for GF40 for 28 days (especially 30% granite fines has given higher values than reference mix GF0).

Interestingly, load-bearing capacity of cylindrical specimens for GF40 has drastically come down when compared to GF30 by 67%, which clearly reveals that a 30% of replacement of natural sand by granite fines yields a modulus of elasticity value greater than the reference mix GF0 and that of codalmodulus of elasticity value of 22,500 (5,000 $\sqrt{f_{ck}}$).

From the study it is clear that at 30% of replacement of natural sand by granites yield higher modulus of elasticity values. The test results are supported by the study of Kanamalai et al.

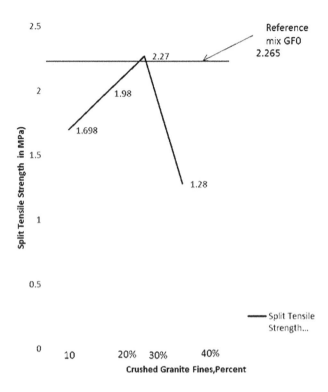

FIGURE 25.1 Split tensile strength values for 0%, 10%, 20%, 30% and 40% replacement of natural sand with granite fines.

25.4.4 X-Ray Powder Diffraction Analysis

X-ray diffraction (XRD) analysis has been used for the fingerprint characterization of pozzolanic material and for the determination of the crystal structures. The basis of the diffraction phenomenon is that the wavelength of the incident radiation (i.e., X-rays) is in the order of magnitude of the inter atomic distance in the crystalline solids. Therefore, when an X-ray beam falls on a crystalline material, it will be diffracted by the incidence angle given by Bragg's law:

$$2d \sin\theta = n\lambda \tag{25.3}$$

Where d is the interplanar spacing, λ is the wavelength of the X-ray, n is an integer and θ is the diffraction angle. For different 2θ values, the interplanar spacing d is calculated, the 2θ values used are those that give the highest intensities (i.e.,peaks) when the sample, in the form of a powder, is scanned. Each crystalline structure is characterized by a univocal set of θ-values that corresponds to their crystal planes, thereby allowing the identification of the different crystallographic phases of the material. In the case of amorphous materials, the scattering of the X-ray beam is diffuse and the XRD pattern shows broad humps rather than sharp peaks. A standard database (JCPDS database) for X-ray power diffraction enables phase identification for a large variety of crystalline phases in a sample. The various figures of XRD analysis illustrate the XRD diffractograms for various investigated samples. The results of the diffraction peaks of the various samples are highly varied indicating that the samples contain variety of crystalline phases. The intensity of XRD peaks of different additives are compared with standard Portland slag cement and ordinary Portland cement samples. The results indicate that the peaks are broadened or slightly decreased or diminished depending up on the additive chemical constituents reactions with $Ca(OH)_2$.

The XRD results of ordinary Portland cement with partial replacement with 20%, 30% by granite fines are presented in Figure 25.3 and Figure 25.4 and for control mix represented in Figure 25.2. XRD results are presented in Table 25.4.

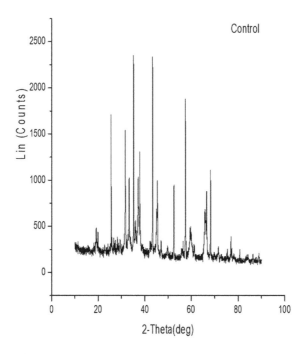

FIGURE 25.2 XRD results for control concrete mix.

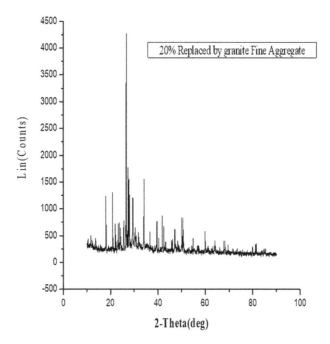

FIGURE 25.3 XRD results for 20% replaced by granite fine aggregate.

From the XRD results presented in Table 25.4, it is shown that the hexagonal structure is changed to orthorhombic or anorthic, with the inclusion of granite fine aggregate with replacement of 20%, 30% by weight, and the 2θ angle is around 26° in both the cases, which are identified by comparing the results with standard JCPDS data. From the various results discussed in the previous sections, it can be understood that the increase in the compressive strengths up to certain level (ie., 20%) and the decrease in the

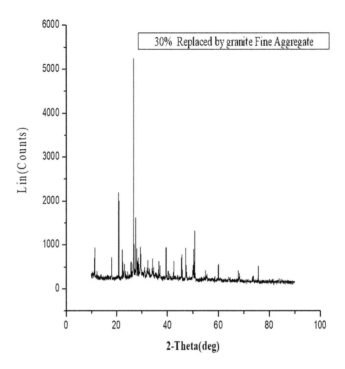

FIGURE 25.4 XRD results for 30% replaced by granite fine aggregate.

TABLE 25.4

XRD Patterns

S.No.	Cement+ Replacement material	2θ Angle (degrees)	Structure	Wavelength (A0)	Possible Compound	JCPDS Number
1.	OPC	35.164	Hexagonal	1.54060	$CaAl_4Si_2O_{11}$ Calcium Aluminum Silicate	51-0093
2.	OPC+20% replacement of granite fine aggregate	26.6864	Orthorhombic	1.54060	$Al_2(SiO_4)O$ Aluminum Silicate	89-0890
3.	OPC+30% replacement of granite fine aggregate	26.6864	Anorthic	1.54060	$Na(AlSi_3O_8)$ Calcium Aluminum Oxide	89-6429

compressive strengths up to certain level (ie., 30%) of replacement and enhancement in the durability properties, are occurring because of the partial replacement of granite fine aggregate in coarse aggregate. The changes in the structure and formation of various compounds and filling of the pore spaces with C-S-H further accelerates the pozzolanic actions which are matching with the results presented in XRD analysis.

The XRD patterns in this study were obtained as shown in above figures and possible compounds are shown in Table 25.4.

25.5 Conclusion

From the test results it is clearly evident that granite powder as a partial replacement of sand proves to be beneficial when mechanical properties of strength are considered. Concrete mix with 30% addition of granite fines proves to be better when compared with other proportions. The following conclusions can be arrived after comparing GF30 with the reference concrete mix GF0.

1. Split tensile strength, and modulus of elasticity, particularly in all the ages higher than reference mix (GF0), there was an increase in strength as the days of curing increases.
2. Split tensile strength, and modulus of elasticity, at 30% replacement of sand with crushed granite fines resulted in a peak value, higher than reference mix (GF0).

REFERENCES

Bahoria, B.V., Parbat, D.K., and Naganaik, P.B.,: Replacement of natural sand in concrete by waste products: A state of art, *Journal of Environment Research and development*, Vol. 17, April–June 2013, pp. 1651–1656.

BaharDemitrel: The effect of the using waste marble dust as fine sand on the mechanical properties of concrete, *International Journal of Physical Sciences*, Vol. 5, August 2010, pp. 1372–1380.

Hameed, M.S. and Sekar, A.S.S.: Properties of green concrete containing quarry rock dust and marble sludge powder as fine aggregate, *ARPN Journal of Engineering and Applied Sciences*, Vol. 4, 2009, pp. 83–89.

IS 2386 (part-1): 1963: Methods of test for aggregates for concrete, part I-particle size and shape, Indian standards Institution, New Delhi.

IS 2386 (Part-I): 1963: Methods of test for aggregates for concrete, Part-III-Specific gravity, density, voids, absorption and buckling, Indian Standards Institution, New Delhi.

IS 456: 1978: Code of practice for plain and reinforced concrete, Indian Standards Institution, New Delhi.

IS 456: 2000: Code of practice for plain and reinforced concrete, Indian Standards Institution, New Delhi.

IS 516: 1999: Methods of test for strength of concrete.

IS 10262: 2009, Recommended guidelines for concrete mix design, Indian Standard Institution, New Delhi.

Kanmalai Williams, C., Partheeban, P., and Felix Kala, T..: Mechanical Properties of High Performance concrete incorporating granite Powder as fine aggregate, *International Journal on design and Mnufacturing technologies*, Vol. 2, July 2008, pp. 67–73

Olawvyi, B.J., Olusula, K.O., Babafemi, A.J.: Influence of curing age and mix composition on compressive strength of volcanic Ash blended cement laterized concrete, *Civil Engineering Dimension-Journal of civil Engineering Science and application*, Vol. 14, 2012, pp. 84–91.

Prince Arulraj, G., Adin, A., Suresh Kannan, M.J.: Granite powder, concrete, IRACST-engineering science and technology: *An International Journal*, Vol. 3, February 2013, pp. 193–198.

Raman, S.N., Safiuddin, M., Zain, M.F.M.: Non-Destructive Evaluation of flowing concretes incorporating Quarry waste, *Asian Journal of Civil Engineering (Building and Housing)*, Vol. 8, 2007, pp. 597–614.

Safiuddin, M.D., Raman, S.N., Zain, M.F.M.: Utilization of Quarry waste fine aggregate in concrete mixtures, *Journal of Applied Sciences Research*, Insinet Publications, Vol. 9, 2007, pp. 202–208.

Shirulea, P.A., Ataur Rahman, B and Rakesh, D.G.: Partial replacement of cement with marble dust powder, *International Journal of Advanced Engineering Research and Studies*, Vol. 1, April–June 2012, pp. 175–177.

Uchikawa, H.S., HaneharaHirao, H.: Influence of microstructure on the physical properties of concrete prepared by substituting mineral powder for part of fine aggregate, *Cement and Concrete Research*, 1996, pp. 101–111.

26

Eco-Friendly Utilization of Industrial Sludge as a Building Material: A Study of Steel Industries in the Tarapur Region, Maharashtra

Tarang Jobanputra
Charotar University of Science and Technology, Changa, Gujarat

Vaidik Gajera
U. V. Patel College of Engineering, Ganpat University, Ahmedabad, Gujarat

Gaurav Kapse
Charotar University of Science and Technology, Changa, Gujarat

CONTENTS

26.1 Introduction 219
26.2 Study Area 220
26.3 Materials and Methodology 220
    26.3.1 Materials 221
    26.3.2 Methodology 221
26.4 Results and Analysis 224
26.5 Conclusion 226
References 227

26.1 Introduction

In the 21st Century, the point where we have reached a stage that clamors for protecting the planet has proliferated substantially. Humanity, in the form of droughts, sea level rise, and global warming has been facing such unprecedented consequences of industrial activities and greenhouse gas emissions that we had been emitting for centuries. Waste production has nowadays become a conspicuous outcome of the developmental activities, as industries have become an inevitable part of modern society [1]. Increasing pressure on secured landfill sites and stringent environmental laws [2] have made it necessary for companies to follow waste management practices. In India, there are a total of 41,523 hazardous waste generating industries in the country, generating about 7.90 million tonnes of hazardous waste annually, out of which 42% is landfillable, 50.4% is recyclable, and 7.6% is incinerable, but about 63% of waste is being dumped in the landfills, and thus creating ecological problems [3].

Presently, India is currently the world's third largest producer of crude steel, which contributes to ~2% of the country's GDP and employees ~2,500,000 people. As per the current practice, for every one tonne of steel produced, 1.2 tonnes of solid waste is produced in India, compared to 0.55 tonnes in Western countries [4]. Thereby, recent advancements have been given attention in order to maximize the eco-efficiency, i.e., by maximizing the physical recycling and minimizing the environmental burden [5]. It has become a prominent trend, nowadays, to use the industrial by-products, coke oven sludge, distillation residue, mixed waste solvent, spent carbon and many other waste materials, as partial fuel in the cement kiln. Plastic waste can also be recycled and can be used for road making and co-processing in a cement plant, and chemical gypsum can be used for cement production [6].

In the race to reuse and recycling, much research had been done to reuse the steel industrial waste in construction, since the chemical components of many building materials are known to be compatible with iron [7, 8, 9]. Out of which, slag is the most widely and effectively reused waste material (100% of the generated) [4]. Hence, the usage of the industrial wastes in concrete for building construction could be economically beneficial since the material cost can be reduced [10]. Due to the high strength and durability of steel industrial slag, sometimes, considered as superior to rock material [11], and it also increases the resistance to deflection and vertical strain in case of beams [12]. It had also been observed to capture and store atmospheric carbon from steel and construction industrial wastes, in order to reduce the net anthropogenic CO_2 emissions and subsequent climate change [13]. That too, reuse of industrial and urban wastes in clay raw materials were found useful in brick manufacturing, as the correct disposal of bulky and polluting materials as well as the reduction in the cost of raw materials, energy savings and laborers [14].

In the present work, an intimate study had been done to reuse the steel industrial waste by taking into account the paradigm of circular economy and by studying several components of the amount of waste generated, properties (both physical and chemical), research and its utilization in construction works, so as to reduce the load on the secured landfill sites of Taloja and subsequent cost associated. An attempt has been made by making solidified blocks along with other materials such as lime, cement, grit dust and aggregates that can be used for partition walls and other construction purposes, where a no-load-bearing structure is required.

26.2 Study Area

Tarapur is located in the Thane district at 17°42′N 75°28′E, 17.7°N 75.47°E, having one of the largest industrial estates in the state of Maharashtra, known as MIDC Tarapur. It possesses 1,548 large and medium-scale and 18,480 small-scale industries. The major industries include iron and steel, electronics, drugs, textiles, and chemicals [15].

The present study is conducted for the steel industries at the Tarapur Industrial Area since iron and steel industries incorporate a significant chunk in the production sector and subsequently, results in a large quantity of hazardous waste generated. The research work focuses mainly on the utilization of two types of sludge, which are being generated from the industrial processes at a large scale, namely:

1. Effluent Treatment Plant sludge (ETP sludge)
2. Bonder sludge

The steel industrial intricate processes result in the hazardous nature of the sludge generated, which further needs to be disposed off in secured landfill sites of Taloja dumping ground. The tentative data available from the site visit concluded that an individual steel industry generates more than 100 tonnes (±5 tonnes) per month of ETP sludge and more than 15 tonnes (±3 tonnes) per month of Bonder sludge within the area.

26.3 Materials and Methodology

For the purpose of making blocks/bricks, apart from two types of hazardous waste, cement and lime for the binding purpose and grit dust and aggregates for providing bulk to the blocks/bricks were used. Part replacement method was used by optimizing various materials one-by-one and to arrive at a final engineered design.

26.3.1 Materials

The materials used in the study include:

Hazardous Sludge: Two type of hazardous waste sludge are being generated primarily on a large scale from the different plants of steel industries, namely ETP/Red sludge and Pickling/Bonder sludge. As the name suggests, the former is being generated by the effluent treatment plant of the industry and the latter from the Pickling Department during the phosphate process. For the purpose of research and testing, both sludges were stored in 50 litre air-tight containers [18] in order to avoid further chemical reaction with the atmosphere, throughout the research work.

Slaked Lime: Lime is calcium-containing inorganic mineral in which carbonates, oxides, and hydroxides predominate. The sole purpose of incorporating lime is due to its excellent plasticity properties, resistance to moisture, and less shrinkage on drying. In the present study, slaked lime available from the local region of Tarapur is added in various proportions with sludge, to check its compatibility and to make the final block design economically feasible, thereby using a cement-like material. Slaked lime was stored in air-tight containers at 1m above the ground in an insulated room in order to avoid its reaction with the atmospheric carbon dioxide.

Grit Dust: During the crushing process of rocks, a by-product in the form of waste is generated is called quarry dust. It is generally used as aggregates (fine) in concrete, and also as a filler material for finished bituminous road surfacing. In current work, grit dust available from the local region of Tarapur was used in order to provide bulk to the mixture and maintain the economic feasibility of the design. The material was compatible with IS 383-1970 [20] and belonged to Zone-III, with 80% of grains with size less than 1.18mm and less than 600μm. The specific gravity was 2.55, and fineness modulus was 2.60. The material used in the study was free from impurities and stored in air-tight containers in order to avoid further chemical reaction with the atmosphere and ground throughout the research work.

Aggregates: Construction aggregate, is a broad category of coarse to medium-grained particulate material used in construction, in general, due to its uniform properties compared to other soils, which includes high hydraulic conductivity value and strength. In the present study, aggregates of size less than 2.5mm and compatible with IS 383 (1970) [20] and were tested as per IS 2386 (1963) [21], mainly used to provide bulk. The main aim of the usage of aggregates was in order to check its compatibility with sludge and other materials. A size of 2.5mm was preferred to avoid honeycombing of blocks, and the specific gravity and fineness modulus was 2.69 and 2.80, respectively.

Cement: It is a mixture of compounds made by burning limestone and clay together at very high temperatures ranging from 1400°C to 1600°C. It was mainly used in the current project because of its high compressive strength, durability, and flexible configuration. In the present study, 53 Grade OPC Cement available from the local region of Tarapur and compatible with IS 12269 (2013) [22] was used. Cement was incorporated in the study in order to check its compatibility with sludge along with lime. Cement bags were stored in an insulated room at 1m above the ground level in order to avoid the reaction with the moisture in the atmosphere and ground.

26.3.2 Methodology

The scope of the study was to utilize the maximum possible percentage of hazardous waste generated by the industry. The study methodology was divided into two parts:

1. Characterization of sludge: As there was very little information available on the sludge, the properties of both the sludge were checked for its physical and chemical properties. Physical properties were analyzed as per IS 2720 (1980) [23] in terms of its specific gravity, LOD and LOI as per method 1684 (USEPA) [19], and initial and final setting times as per IS 4031 (1988) [27]. Chemical analysis in terms of pH, chlorides and sulphates by SW 846 method [17], performed at the Environmental Engineering Laboratory, Department of Civil Engineering, CSPIT,

CHARUSAT, and the XRF analysis was done at Sophisticated Instrumentation Centre for Applied Research and Testing (SICART), Vallabh-Vidhyanagar [28–31].

2. Experimental Trials for Mix Proportions: For the purpose of testing the material, the standard procedure of batching, mixing, placing, curing, and testing was adopted as per IS 516 (1959) [28]. In batching, weight batching was adopted. To reduce the quantity of material for the test, standard moulds of size 70mm x 70mm x 70mm were taken, and hand mixing was adopted. After mixing, the material was filled in the mould in layers of 25mm and each layer was compacted by a 16mm diameter tamping rod [27] and then on a vibrating table. Since, the binding materials of the mix included cement and lime, both methods, i.e., Gunny bag curing and Air setting, were adopted [24]. Finally, at the end of 7 days, the cubes were tested in the Standard Compression Testing Machine and results were obtained.

The experimental trials adopted based on part replacement sequence were:

Trial 1 to 4 - Optimization of ETP sludge: To check the compatibility of ETP sludge with lime, it was mixed in proportions as 0.2, 0.3, 0.4 and 0.5 parts to 1 part of lime. The water content was kept as 30% of the total weight of 600g of the mixture so as to make a homogenous mixture to form a block. A total of 12 blocks (3 for each proportion) were prepared and then were cured and tested as per the standard procedure mentioned previously in 2.2.1. Table 26.1 represents the values of ETP sludge and lime in proportions, and the compressive strength of the average of three for each proportion was taken as final results (Figure 26.1).

Trial 5 to 8 - Optimization of Bonder sludge: To check the compatibility of Bonder sludge along with the best mixture of ETP + lime (in terms of strength), it was varied as 0.5, 0.6, 0.7 and 0.8 to the best proportion among trial 1–4. The same procedure of testing for trial 1-4 was followed. Table 26.2 represents the value of lime, ETP, and Bonder sludge in proportions and the compressive strength of the average of three for each proportion was taken (Figure 26.1).

Trial 9 to 12 - Optimization of aggregates: To impart bulk and strength to the blocks, aggregates were added in the proportion of 0.5, 1.0, 1.5, and 2.0 to the best mixture from trial 5–8, the same procedure adopted for trial 1–4 was followed. Table 26.3 represents the value of lime, ETP, Bonder sludge, and aggregates in proportions and the compressive strength of the average of three for each proportion was taken (Figure 26.1).

Trial 13 to 16 - Optimization of grit dust: To check the compatibility and to provide bulk to the blocks, waste grit dust was used in the proportion of 0.5, 1.0, 1.5, and 2.0 to the best mixture from trial 5–8, the same procedure adopted in trial 1–4 was followed. Table 26.4 represents the value of lime, ETP, Bonder sludge, and grit dust in the proportions and the compressive strength of the average of three for each proportion was taken.

Trial 17 to 22 - Optimization of Cement: To impart strength to the blocks, cement at last was introduced in the mixture, so as to check its compatibility with sludge and other materials binding

TABLE 26.1

Optimization of ETP Sludge

Trial	Mix Proportion (L : E) Lime	Mix Proportion (L : E) ETP	Proportion of total 600g (g) Lime	Proportion of total 600g (g) ETP
1	1	0.2	480	120
2	1	0.3	420	180
3	1	0.4	360	240
4	1	0.5	300	300

Eco-Friendly Utilization of Industrial Sludge

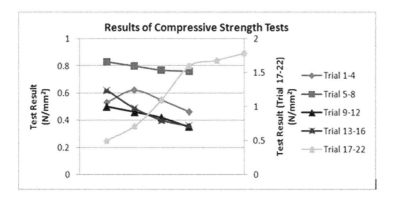

FIGURE 26.1 Compressive strength of various trials at 7 days.

TABLE 26.2
Optimization of Bonder Sludge

Trial	Proportion (L : ES : BS)			Proportion of total 600g (g)		
	Lime	ETP	Bonder	Lime	ETP	Bonder
5	1	0.3	0.5	334	100	166
6	1	0.3	0.6	315	95	190
7	1	0.3	0.7	300	90	210
8	1	0.3	0.8	286	86	228

TABLE 26.3
Optimization of Aggregates

Trial	Proportion (L : ES : BS: A)				Proportion of total 600g (g)			
	Lime	Sludge ETP	Sludge Bonder	Agg.	Lime	Sludge ETP	Sludge Bonder	Agg.
9	1	0.3	0.5	0.5	305	90	152	152
10	1	0.3	0.5	1	250	75	125	250
11	1	0.3	0.5	1.5	212	64	105	320
12	1	0.3	0.5	2	184	56	93	369

TABLE 26.4
Optimization of Grit Dust

Trial	Proportion (L : ES : BS: GD)				Proportion of total 600g (g)			
	Lime	Sludge ETP	Sludge Bonder	Grit Dust	Lime	Sludge ETP	Sludge Bonder	Grit Dust
13	1	0.3	0.5	0.5	305	90	152	152
14	1	0.3	0.5	1	250	75	125	250
15	1	0.3	0.5	1.5	212	64	105	320
16	1	0.3	0.5	2	184	56	93	369

TABLE 26.5
Optimization of Binding Material

	Proportion				Proportion of total 6,800 (g)				
	Binding	Sludge					Sludge		
Trial	Material (L + C)	ETP	B	Grit	Lime	Cement	ETP	B	Grit
17	1.0	0.3	0.5	0.5	2910 (100%)	0 (0%)	875	1475	1515
18	1.0	0.3	0.5	0.5	2330 (80%)	580 (20%)	875	1475	1515
19	1.0	0.3	0.5	0.5	1750 (60%)	1160 (40%)	875	1475	1515
20	1.0	0.3	0.5	0.5	1160 (40%)	1750 (60%)	875	1475	1515
21	1.0	0.3	0.5	0.5	580 (20%)	2330 (80%)	875	1475	1515
22	1.0	0.3	0.5	0.5	0 (0%)	2910 (100%)	875	1475	1515

materials were kept in proportion of 1 to the best proportion from trial 13 to 16. Cement was used as a partial replacement for lime in the mixture as 100%, 80%, 60%, 40%, 20%, and 0% for the total portion of binding materials. For the final check, and to arrive at definite results, a total of 36 blocks (3 for each proportion) were prepared of standard size of 150mm × 150mm × 150mm and were then cured and tested at the end of 7 days and 28 days as per the standard procedure. The water content was kept as 30% of the total weight of 6800g per block of the mixture so as to make a homogenous mixture. Table 26.5 represents the value of lime, ETP sludge, Bonder sludge, grit dust and cement in various proportions and the compressive strength of the average of three for each proportion was taken as final results (Figure 26.1).

26.4 Results and Analysis

By preliminary analysis, it was observed that, based on smell, ETP sludge might be alkaline and bonder sludge might be acidic in nature, and the brownish color of ETP sludge indicated the presence of ferrous or ferric compounds. Moreover, the results of the physical, chemical and XRF analysis are tabulated in Table 26.6 and Table 26.7.

The physical analysis revealed that moisture content of around $2/3^{rd}$ and $1/4^{th}$ of its weight was there in ETP and Bonder sludge, respectively. Also, it was observed that no self-binding remains at high

TABLE 26.6
Physical and Chemical Analysis Results

				Results	
Sr. No.	Parameters	Unit	Test Method	ETP Sludge	Bonder Sludge
1	pH	-	SW 846	11.6	4.16
2	Specific Gravity	-	ASTM D 5057-90	1.92	1.92
3	Loss on Drying @ 105°C	%	APHA 22nd Ed., 2012, 2520	63.69	22.3
4	Loss on Ignition @ 550°C	%	APHA 22nd Ed., 2012, 2520	26.27	18.6
5	Chlorides	mg/liter	Argentometric	200	74
6	Sulphates	mg/liter	Turbidimetric	331.7	120

TABLE 26.7

XRF Analysis Results

Sr. No.	Metal	Concentration (%) ETP Sludge	Bonder Sludge
1	Na	0.161	0.335
2	Mg	0.367	-
3	Al	0.171	-
4	Si	2.515	0.872
5	S	3.432	0.508
6	Ca	**30.714**	0.335
7	Mn	0.387	0.16
8	Fe	**38.154**	**63.584**
9	Ni	0.124	-
10	Cu	3.988	0.102
11	Zn	4.532	**25.369**
12	Sr	0.14	-
13	Ru	0.359	0.253
14	Sn	0.194	-
15	Pb	12.925	-
16	P	-	8.04

temperature. From the chemical analysis, it was observed that the sludge contains significant amount of sulphate (SO_4^{-2}), which gave an indication to have better binding properties with lime (CaO) and cement. The XRF analysis revealed that for ETP sludge high concentration of Fe and Pb were observed due to steel processing operations in the industry, and for Bonder sludge high concentration of Fe and Zn observed may be due to the use of chemicals, coating materials and steel processing operations in the industry. Both chemical and XRF analysis bolstered the claim to use steel industrial sludge, as it contained high concentration of iron compound, which has a better bonding capacity with cement compounds. The chemical analysis also underscored a significant concentration of hazardous lead (Pb) in ETP sludge, which hinted of a potential to be reused.

With this background of the chemical composition, experimental trials were fixed and by part-replacement sequence method, optimization of materials was performed and testing of their respective compressive strength was carried out. For each trial (mentioned from Table 26.1 to 26.5) the best proportion was selected, and new materials were introduced for each trial. The results of the compressive strength of each trial can be further interpreted from the graphs.

From Figure 26.1, aggregates used (trial 9–12) in turn had low or inverse effect on the strength of blocks and hence avoided. Further, waste grit dust was used instead of aggregates [16]. After the rigorous process of trial and testing, the final proportions were fixed from the results of trial 17–22. Further, the actual strength was predicted by the formula mentioned below and was compared with the actual strength at 28 days in the equation:

$$\sigma_{28} = 1.4\,\sigma_7 + 150$$

where, σ = Compressive strength in pounds/sq. inch. [32]

From the data in Table 26.6 and Figure 26.1, it is clear that trial–22 (100% cement and 0% lime) in the mix shows a significant strength of around 3.58N/mm^2 at the end of 28 days. Though, it can be seen that there is no significant increase in compressive strength from trial–20 to trial–22. Hence, considering the economical aspect, trial–20 was considered as the optimal solution for the combination of various materials for production of the blocks.

TABLE 26.8

Results of the Compressive Strength for the Final Trial

Trial	Compressive Strength @7 Days (N/mm²)	Compressive Strength @28 Days as per formula (N/mm²)	Actual Compressive Strength @28 Days (N/mm²)
17	0.5	1.73	1.65
18	0.71	2.00	2.1
19	1.1	2.57	2.63
20	1.55	3.20	3.29
21	1.68	3.38	3.4
22	1.79	3.54	3.58

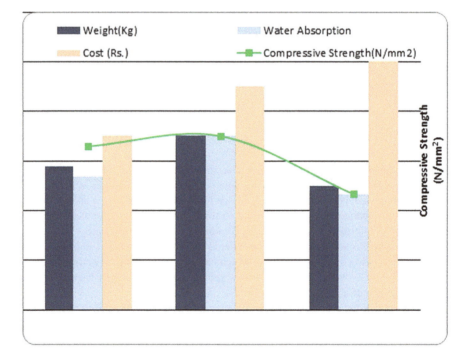

FIGURE 26.2 Graph showing comparisons among blocks, conventional bricks, and bricks, used in Tarapur Region.

After finalizing a set of proportions, the density of the blocks was found to be 1.79gm/cm³, which is less than 2.92gm/cm³ (i.e., for the conventional brick.). Water absorption was found to be 2.68%, which is less than 20% (IS 3495 (1992)) [30] and bio-assay acute toxicity test result showed that fishes stayed alive for a period of more than 72 hours. (IS 6582 (1971)) [31].

The final material/mass balance calculation, inclusive of handling, transportation, and labor cost, revealed that cost per block would be Rupees 3.5/-. Further comparison of the blocks with conventional bricks and the bricks that are used in the Tarapur region is given in Figure 26.2.

26.5 Conclusion

Studies reveal that these blocks made with steel industrial sludge are lighter in weight (at least by 60% as compared to brickwork), which will lead to more economical frame structure, and they can be used in non-structural elements. The experimental results indicated that there is an increase in strength with decrease in lime content and increasing cement content indicates that the sludge material is more suitable for cement as a binder as compared to that of lime, but for a cost-effective solution, lime can be used. The maximum strength obtained with 44% of binding material (cement + lime), 22% grit dust, and 35% of hazardous material (ETP + Bonder) will be considered the best suitable proportion for the blocks. The authors also recommend future analysis, which still can be done to analyze the leachate, durability and shrinkage, soil stabilization properties and plasticity characteristics of the material.

REFERENCES

1. V. Mishra, S. Pandey, Hazardous waste, impact on health and environment for development of better waste management strategies in future in India, *Environment International*, Vol. 31, Issue 3, 417 (2005).
2. Ministry of Environment, Forest and Climate Change, Gazette of India, Hazardous and other wastes (management and transboundary movement) rules, Extraordinary, Part II, Section 3, Sub-Section (i), New Delhi (2016) https://cpcb.nic.in/displaypdf.php?id=aHdtZC9IV01fUnVsZXNfMjAxNi5wZGY.
3. Central Pollution Control Board, Hazardous Waste management Division, Existing scenario of hazardous waste management in India, Delhi, India (2008) http://cpcbbrms.nic.in/writeread/HW%20Management.pdf.
4. Ministry of Steel, Government of India (2017-18), Annual Report. https://steel.gov.in/sites/default/files/annual_Report_E_07March2018.pdf
5. C. Hagelüken, Improving metal returns and eco-efficiency in electronics recycling - a holistic approach for interface optimisation between pre-processing and integrated metals smelting and refining, Proceedings of the 2006 IEEE International Symposium on Electronics & the Environment, 8 (2006), Scottsdale, AZ, USA.
6. B. Sengupta, Workshop on "Hazardous waste management: waste to wealth", CEE (2016).
7. C. Khajuria, R. Siddique, M. Kumar (2013). http://hdl.handle.net/10266/2210
8. H. Yi, G. Xu, H. Cheng, J. Wang, Y. Wan, H. Chen, An Overview of Utilization of Steel Slag, *Procedia Environmental Sciences*, Vol. 16, 791–801 (2012).
9. S. Malhotra, S. Tehri, Building Materials from granulated blast furnace slag – Some new prospects, *Indian Journal of Engineering and Material Sciences*, Vol. 2, 80 (1995).
10. A. Ananthi, J. Karthikeyan, Properties of industrial slag as fine aggregate in concrete, *International Journal of Concrete Technology*, Vol. 1, Issue 1 (2015).
11. M. Tossavainen, E. Forssberg, Leaching behaviour of rock material and slag used in road construction – a mineralogical interpretation, *Steel Research*, Vol. 71, Issue, 11, 442–448 (2000).
12. S. Yildirim, M. Prezzi, *Advances in Civil Engineering*, Hindawi Publishing Corporation (2011), doi:10.1155/2011/463638
13. Stolaroff, G. Lowry, D. Keith, Using CaO- and MgO-rich industrial waste streams for carbon sequestration, *Energy Conversion and Management* 46, 687–699, (2005).
14. M. Dondi, M. Marsigli, B. Fabbri, Recycling of urban and industrial wastes in brick production: a review, *Tile & Brick International* 13(3), 218 (1997).
15. Maharashtra Pollution Control Board, Action plan for Tarapur industrial area, Tarapur, Maharashtra, India (2010) http://mpcb.gov.in/CEPI/pdf/Action%20Plan%20CEPI-Tarapur.pdf.
16. C. Hanumantha, *Advances in Civil Engineering*, Hindawi Publishing Corporation, 1742769, 5, (2016).
17. U.S. Environmental Protection Agency, *Test Methods for Evaluating Solid Waste, Physical/Chemical Methods*, EPA Publication SW-846, Third edition, Boston, USA (2015).
18. Central Pollution Control Board, Ministry of Environment, Forest & Climate Change, Manual on sampling, analysis, and characterization of hazardous wastes, Government of India, Delhi, India. (2013) https://cpcb.nic.in/displaypdf.php?id=aHdtZC8xOS5wZGY.
19. Total, Fixed and Volatile Solids in Water, Solids and Bio-Solids, USEPA Methods 1684, 11 (2001), U.S. Environmental Protection Agency Office of Water Office of Science and Technology Engineering and Analysis Division (4303) 1200 Pennsylvania Ave. NW Washington, DC 20460- https://settek.com/documents/EPA-Methods/PDF/EPA-Method-1684.pdf
20. IS 383 Indian Standard Specification for Coarse and Fine Aggregates from Natural Sources for Concrete (1970), *Bureau of Indian Standards*, New Delhi, India.

21. IS 2386 Indian Standard Methods of Test for Aggregates for Concrete (1963), *Bureau of Indian Standards*, New Delhi, India.
22. IS 12269 Indian Standard Specification for Ordinary Portland Cement, 53 Grade (2013), *Bureau of Indian Standards*, New Delhi, India.
23. IS 2720 Indian Standard Methods of Test for Soils (1980), *Bureau of Indian Standards*, New Delhi, India.
24. IS 1727 Indian Standard Methods of Test For Pozzolanic Materials (1967), Bureau of Indian Standards, New Delhi, India.
25. IS: 2720-3-1 Indian Standard Methods of Test for Soils, Part 3: Determination Of Specific Gravity, Section 1: Fine Grained Soils (1980), *Bureau of Indian Standards*, New Delhi, India.
26. IS: 2720-3-2 Methods Of Test For Soils, Part 3: Determination Of Specific Gravity, Section 2: Fine, Medium And Coarse Grained Soils (1980), *Bureau of Indian Standards*, New Delhi, India.
27. IS 4031 Indian Standard Methods of Physical Tests for Hydraulic Cement (1988), *Bureau of Indian Standards*, New Delhi, India.
28. IS 516 Indian Standard Methods of Tests for Strength of Concrete (1959), *Bureau of Indian Standards*, New Delhi, India.
29. IS 1199 Indian Standard Methods for Sampling and Analysis of Concrete (1959), *Bureau of Indian Standards*, New Delhi, India.
30. IS 3495 Indian Standard Methods Tests of Burnt Clay Building Bricks (1992), *Bureau of Indian Standards*, New Delhi, India.
31. IS 6582 Bio-Assay Methods for Evaluating Acute Toxicity of Industrial Effluents and Waste Waters (1971).
32. M. Shetty, *Concrete Technology: Theory and Practice*, 6th Edn. Chand Publication, New Delhi (2013).

Index

A

acid resistance, 176
aggregate, 185–188
air pollution, 185
Alkaliphilic bacteria, 120–121, 127
alternative, 209
alumina, 177, 182
ambient, 182
amorphous phase, 64
anomalous, 64
aquatic life, 185
argemone mexicana methyl esters(amme), 96–97
artificial aggregates, 185, 187, 189
ash-based, 175
ash-geogrid interface, 155, 157
Atterberg limits, 119, 121–122, 129, 133
    test, 131–132

B

bagasse ash, 153–155, 158–159
bearing capacity, 113–117
BIM-enhanced, 191
bio-cloggation, 127
biodegradability, 55
biodiesel, 95–101
bio-mining, 74
blast furnace, 65
blasting effects, 185
Bonder sludge, 220, 222–225
bottom ash, 153–155, 156–158
brake thermal efficiency, 96
bricks, 220, 226
bulk density, 186
by-product, 62, 87, 103

C

calcium silicate hydrate (c-s-h) gel, 180
California bearing ratio, 131–132, 134–135
Calorific value, 97, 99
C&D waste, *see* construction, and demolition waste
capsularies, 200
carbon dioxide emission, 87
carbon dioxide gas (CO_2), 12, 87, 129, 165, 175, 221
carbon-footprint, 18
cast, 210, 212
catastrophic, 7
CBR values, 141–142
cement, 185–186, 188–189
    concrete, 77–78, 88, 90, 93
    production, 175
cement replacement materials (CRM), 15

chemical, 194, 200, 202
    compositions, 177–178
clay brick, 89–90
clean, 191, 193, 195, 197
coarse, 177, 185–187, 189
    aggregate, 165, 167–169
codalmodulus, 213
cohesion, 146–147
cohesive non-swelling, 145
compact laser scanners, 194
composite, 185, 188, 200
compression index, 145–147, 150–151
compressive, 176
    strength, 61, 78, 80, 108, 151, 165, 172, 204
concrete, 179, 185, 188–189, 199–201, 203, 205
construction, 192–197
    and demolition waste, 3–4, 191, 193, 195, 197
    material, 67, 77–78, 87, 103, 108
contaminating, 70, 73
cost-effective, 119, 127
crusher dust, 114–116
curing period, 124, 127
    time, 65
cylinder block, 97

D

deconstruction, 191, 193
de-gumming, 96
demolding, 80
density, 165–168, 170, 172
digestion, 70, 103
dismantling, 191
disposal, 69, 73–74, 85
dry density, 154–155
dumpsite, 69, 74
durability, 185
    test, 108

E

earthquake, 4
eco-friendly, 54
economy, 7, 11
    analysis, 139, 143
ecosystem, 87, 103
efficiency, 195–196
efflorescence, 58, 175–176, 178, 181
energy dispersive X-ray analysis (EDX), 34–35
environment, 185, 192, 197
environmental, 175
    impact, 87, 104
    protection agency (EPA), 4, 53
esterification, 96

Index

ETP sludge, 220, 222, 224–225
European waste framework directive, 129
e-waste (electronic waste), 12
exothermic reaction, 63
experimental, 211, 213

F

feldspar, 92
ferric compounds, 224
ferrochrome, 175–177
    ash (FCA), 176
    slag, 176
fertile soil, 70
fertiliser industry, 103
fibre reinforced concrete (FRC), 43
fine aggregates, 165, 167, 169
fineness modules zone, 186
flexible pavement, 139–140, 143
flexural strength, 165–166,169,171–172
fly ash, 37, 40, 55, 61, 88, 104, 154, 165, 170, 175
    -based, 199–201, 203
    -geogrid interface, 155
fossil fuel, 95, 175
free fatty acid, 96
friction efficiency factors, 153, 158–159
fruit wastes, 129–131, 133, 135
    powder, 131–133, 134–135
functionally graded material (FGM), 29–31, 32–35

G

gas cleaning plants, 176
geo-accumulation index, 69–71
geogrid, 114–116, 153–155, 157–159
geopolymer, 61, 67, 165, 169–170, 172, 176
    matrix, 179
geopolymerization, 176–182
geo-solid, 53
geosynthetic, 116
global warming, 219
    potential, 5
grain size, 114–115
granite, 211–217
    fines, 209–210
gravel, 211–212
green buildings, 3
greenhouse effect, 85, 87, 95, 97, 99, 101
greenhouse gas, 8
green technologies, 18
gross domestic product, 12
ground granulated blast furnace slag (GGBS), 30, 32–33, 35, 62, 65, 88, 93, 175
ground water, 70, 74

H

hardened, 199
hazardous, 23, 220–221, 225, 227
high density polyethylene (HDPE), 43, 130

high range water-reducing admixture (HRWRA), 24
homogeneous, 58
hydration, 24
hydrocarbon emission, 100
hydrolyzing, 120

I

impurities, 192–193
industrial, 175–176, 182, 209
    waste, 78, 85, 87–93, 113–114
industry, 199–200, 202, 204
    solid waste, 77, 79, 81, 83, 85
infrastructure, 175
ingredient, 185, 188
inorganic silt, 154
intergovernmental panel on climate change (IPCC), 13
International Energy Agency (IEA), 14
interplanar spacing, 214
investigation, 185–189

J

jute fiber, 199–201, 203–205

L

lamination, 29
landfilling, 5, 53
land-locked, 70
land reclamation, 6
layer thickness reduction, 140
leachate, 227
leadership in energy and environmental design (LEED), 4
life cycle assessment (LCA), 5
lightweight fly ash fine aggregates (LWFA), 38
lime, 176
linear variable differential transducers (LVTD), 114
liquid limit, 139–141, 145–147, 151
loss on ignition (LOI), 63
low-density aggregate, 165–168, 170

M

manufacturing, 175, 177, 182
marble slurry, 77–83
material, 175–179, 181–182, 185–186, 191–197
material sustainability indicators (MSIS), 18
maximum dry density, 131, 133–134, 141, 146–147
mechanical, 199–200
metal leaching, 73–74
microbial induced calcium carbonate precipitation, 119
micro-cracks, 65
microstructure, 65–66
micro studies, 119, 126
mineralogical, 176
model footing test, 113–114, 117
modified proctor compaction test, 140
modulus of elasticity, 210–213, 217
    properties, 209, 211, 213, 215, 217

Index

montmorillonite structure, 146
mortar, 70, 78–81, 83–85, 89–90
municipal solid waste, 69, 71, 73

N

natural coarse aggregate, 166, 169
nickel, 69, 71, 73
noise, 185
non-metallic minerals, 3

O

Olitorius, 200
open-dumping, 129
optical process control/one person company (OPC), 30–32, 34–35
optimum moisture content, 154
Ordinary Portland Cement (OPC), 15, 61
Organisation for Economic Co-operation and Development (OECD), 11

P

paper pulp, 55
particulate emission control unit, 153
pavement, 139–140, 143
permeability, 88, 104, 120
petrochemical, 95
phosphogypsum, 103–109
pH value, 119, 121, 124–125, 127
plasticity characteristics, 121, 124, 127
plasticity index, 139–141
plastic limit, 124, 127, 146–147
pollutants, 191–193
Polyethylene Terephthalate (PET), 15, 18
polypropylene, 44
polyvinyl alcohol (PVA), 15
porosity, 88–89
porous structure, 177
portland cement, 61, 77, 91
pozzolanic material, 6, 41–42, 153–154
primary source, 176
proctor compaction test, 132
   test, 154
pulverized, 62

Q

quarry dust, 185–189
quenching, 62

R

recovery-oriented, 191
recycled aggregate concrete (RAC), 90
recycling, 5, 7
red mud, 88–93
reference technology scenario (RTS), 14
reinforced, 17

cement concrete, 5
reinforcement, 24, 113–114, 116–117
relative density, 113, 116–117
resource, 186, 191–192, 197
   circulation, 4
   conservation, 4
rheological, 199
rice husk ash (RHA), 77, 88, 90, 175
river-beds, 209

S

saline content, 89
sand, 185, 216–217
scanning electron microscopy (SEM), 34–35, 62–63, 65
SCBA, 78, 79, 82–83, 85
sea level rise, 219
sedimentation, 185
segregation, 199
seismic, 5
self-compacting, 28
separation, 192
shear strengths, 153
shrinkage, 221, 227
sintered fly, 165, 167, 171
situ, 179, 181
slag, 23
slump cone test, 40, 80, 178
sodium hydroxide, 176, 178
sodium silicate, 176, 178
soil, 221, 227
   characteristics, 127
   erosion, 104
   stabilization, 129–130, 132
   surface morphology, 119, 121
solid wastes, 175
solution, 225, 227
sorptivity, 48, 50
source, 175–182
specific gravity, 186
spherical, 177
split tensile strength, 165–166, 171–172, 209–215, 217
SPSS and cost modeling, 139–140
strain-hardening cementitious composites (SHCCS), 18
strength, 175–182, 185, 187–189
structural health monitoring (SHM), 19
subgrade soil, 139–141
substantially, 219
sugarcane bagasse ash, 78, 80–83
super-plasticisers, 27
surface, 177, 181
   runoff, 104
sustainability, 191
sustainable, 175
swell index, 146–147
swelling pressure, 145–147, 149, 151
swelling soil, 145

synthesize, 177
synthetic plant nutrients, 106

T

tedious, 54
tensile, 41
　strength, 87, 91–93, 114
thermal power, 37, 140
transferability, 191–192, 195, 199
transportation, 61, 69, 95, 109

U

unconfined compression strength test, 119, 121, 124, 127, 131, 134–135

V

vermiculite, 209, 211

W

waste, 175–176, 182, 192–197, 219–222, 225
　ceramic, 88
　disposal, 175
wastelands, 96
water, 185–186, 188
　absorption, 80, 88–91
wavelength, 214, 216
wet sieve analysis, 140
workability, 165–166, 169–170, 172

X

X-ray diffraction (XRD), 177, 214
　testing, 210, 213

Z

Zypmite, 103–107, 109

Milton Keynes UK
Ingram Content Group UK Ltd.
UKHW052035070324
439068UK00003B/7